Trilobites, Dinosaurs & Moa Bones
THE STORY OF NEW ZEALAND FOSSILS

This is the first popular introduction to New Zealand fossils and where to find them. In lively text and 160 photographs and drawings, Dr Bruce Hayward traces New Zealand's story in fossils.

Not only shells but a surprising range of fossils can be found about our countryside and each has a story to tell about the past. Coconuts and corals indicate warmer times; fossil footprints and traces tell of vanished animals; fossil plants, even the trunks of trees turned to stone, record the changing face of New Zealand.

Trilobites, Dinosaurs and Moa Bones spans the history of our earth from primitive life forms, to sea dragons and giant ammonites, and the fossil plants, fish and bird bones from our more recent past. It is written for the general audience, interested in the outdoors and nature, by a scientist with an established reputation for making complex subjects easy and enjoyable.

1. This 30cm skull of a long-necked plesiosaur (sea dragon) has been painstakingly excavated from 70 million year old (Cretaceous) rocks found in Hawkes Bay. The fossil bones of numerous extinct sea dragons (marine dinosaurs) have been found in New Zealand rocks but the remains of only two land-dwelling dinosaurs have been found so far.

Trilobites, Dinosaurs and Moa Bones

THE STORY OF NEW ZEALAND FOSSILS

by BRUCE HAYWARD

with illustrations
and photographs by
the author

GORDON ELL, The Bush Press, AUCKLAND

Thanks

I am grateful to many people for their kind assistance in providing information, especially Ian Keyes, Hamish Campbell, Roger Cooper, Ian Raine, James Crampton, Tony Edwards, Dave Francis, Percy Strong, Trevor Worthy, Phil Millener, Hugh Morgans and Jocelyn Thornton. The New Zealand Geological Survey kindly allowed me to photograph many specimens from the National fossil collections which they hold. I also thank Roger Cooper, Tony Edwards, Gordon Ell, Dallas Mildenhall, Phil Millener, Ian Raine, John Simes, Graeme Wilson and Trevor Worthy for use of their photographs. I have been fortunate in having a supportive and understanding family during the time-consuming writing of this book at home.

Published 1990 by
THE BUSH PRESS

Designer: Gordon Ell
Typesetting: Jacobsons

Printed in Hong Kong through Bookprint Consultants Ltd, Wellington.

Published and distributed by Bush Press Communications Ltd, P.O. Box 33-029, Takapuna, Auckland 9, New Zealand.

ISBN 0-908608-43-8

Contents

2. This beautifully preserved fossil shell of a small paua, *Haliotis powelli*, is approximately one million years old and was collected from the sandstone cliffs behind Ohope Beach, Bay of Plenty. Fossil seashells are commonly found in mudstone and sandstone rocks ("papa") in many parts of New Zealand.

Foreword

A fossil is any remains or trace of an animal or plant that is preserved in rock. Despite their abundance in rocks in thousands of places in New Zealand, few people have ever found a fossil, possibly because they do not know where or what to look for. Now with this book there is no excuse, as it explains how and why fossils form, what they look like, how they got to where they are now and suggests over 50 places where it is easy to see examples.

New Zealand fossils record the history of life in this part of the world. The early history between 530 and 80 million years ago parallels other southern hemisphere lands, as ancestral New Zealand was part of the coast of the supercontinent Gondwana throughout

3. These exquisite fossil starfish *(Ophryaster novaezelandiae)* were found in 75 million year old (Cretaceous) rocks at Motunau, North Canterbury. Fossil starfish are very rare finds but their relatives, the sea eggs and their spines, are frequently found fossilised in New Zealand rocks.

this time. Like other places we have fossil trilobites, graptolites, sea lilies, ammonites, ancient lampshells and seed ferns to name just a few of the groups illustrated in this book.

From 80 to 55 million years ago, New Zealand split away from Gondwana and from then on many of its plants and animals evolved separately from the rest of the world, although westerly winds and currents introduced migrants from time to time. Terrestrial mammals never reached our shores and in their place a unique and diverse bird fauna evolved, including the world's tallest bird (a giant moa) and largest eagle, with a 3m wing span. Because much of New Zealand has been under the sea in the last 80 million years, we have a very rich fauna of fossil shellfish, sea eggs, sharks' teeth, crabs and many other forms. Many are documented in this book.

4. Fossils, such as this large shell, are the remains of organisms preserved in rock. They occur quite commonly in many parts of New Zealand and can usually be dug out using a hammer and cold chisel.

What is a fossil?

Fossils are often thought of as old bones, found on rare occasions, deeply buried within rocks. This is only partly true for fossils can be common and in places form whole hillsides. The word fossil comes from the Latin "fossilis" meaning anything dug-up. A fossil is defined as any trace or remains of an ancient plant or animal that is preserved in the rocks. Thus while old bones found in the rocks are certainly fossils, there are many other more common kinds of fossils in New Zealand rocks.

Most fossils consist of the preserved hard parts of an organism, such as the shell, teeth, bony skeleton or woody branches. Less common as fossils are traces of the soft parts of organisms, for example the impressions of leaves, jellyfish or worms. A fossil can also be the preserved track or footprint of a passing animal or the burrow of an organism that lived within the mud or sand before it was hardened to rock.

The study of fossils is called paleontology. The earliest paleontologists were the ancient Greeks of the fourth and fifth centuries BC. They recognised the remains of seashells and fish in the rocks from some of their highest mountains. The writings of these Greek scholars, however, were lost for many centuries and in the Middle Ages Europeans had reverted to thinking that fossils were the works of mysterious and unknown forces, while others believed that they were all formed during the great flood of Noah. In early China fossils were first thought to be dragon's teeth and stone birds, but by the eleventh century AD were recognised as the preserved remains of ancient organisms.

The study of fossils was really begun by Linnaeus and George Cuvier in the late eighteenth and early nineteenth centuries, and paleontology as a science has flourished ever since.

In New Zealand fossils were first collected by the pre-European Maori people, who dug out fossil tusk shells and strung them around their necks as ornaments. The first scientific collections of fossils in New Zealand were made by such notable early European explorers as Haast, Heaphy and Hochstetter, around the middle of last century. Today over forty paleontologists are employed in this country in the study of fossils and their application to everyday life.

5. (At left) This fossil lampshell (brachiopod) comes from rocks near Reefton and lived in the sea about 370 million years ago. Fossils similar to these were known to the Chinese as early as the fourth century AD, but they were thought to be the outstretched wings of stone birds that lived in the rocks and flew about in thunderstorms. (Width 7cm)

Where are fossils found?

Fossils can be found in many parts of New Zealand from sea level to the very tops of some of our highest mountains. As they occur within the rocks, fossils can only be seen where rocks are actually visible. There are many natural exposures of rock all around the country – often in seacliffs, river banks, bluffs and hilltop or mountaintop rocky knolls. Some of the best exposures of rocks where fossils can sometimes be found are man-made road or railway cuttings and quarry faces.

Obvious fossils are present in only a few of the millions of rock exposures throughout New Zealand. The chances of finding them is largely governed by the kind of rock that is present. There are three major kinds of rock: igneous, sedimentary and metamorphic.

6. (At right) Many New Zealand rocks contain abundant fossil shellfish, such as these 8 million year old extinct scallops *Mesopeplum*, in a 40cm slab of sandstone from Anaura Bay, north of Gisborne.

Sedimentary rocks

Most fossils occur in sedimentary rocks. These form from sediment, such as mud, sand and gravel, that accumulated in layers on the floor of the sea or sometimes in lakes, swamps, river plains and as sand dunes. Most sedimentary rocks can be recognised by their layered appearance which was formed by the progressive accumulation of layer upon layer of sediment. If the kind of sediment (composition, grain size) being deposited remains constant over a long period then no layering will be apparent. In most places however, the kind of sediment varies, often coming in pulses with storms, floods, submarine slides and changing currents, producing distinctive layering from different types of sediment.

7. Fossils, like this 10cm crab from the Wairarapa, often blend in with the background rock and are only seen by observant passers-by. This crab *(Tumidocarcinus giganteus)* was buried and fossilised approximately 10 million years ago.

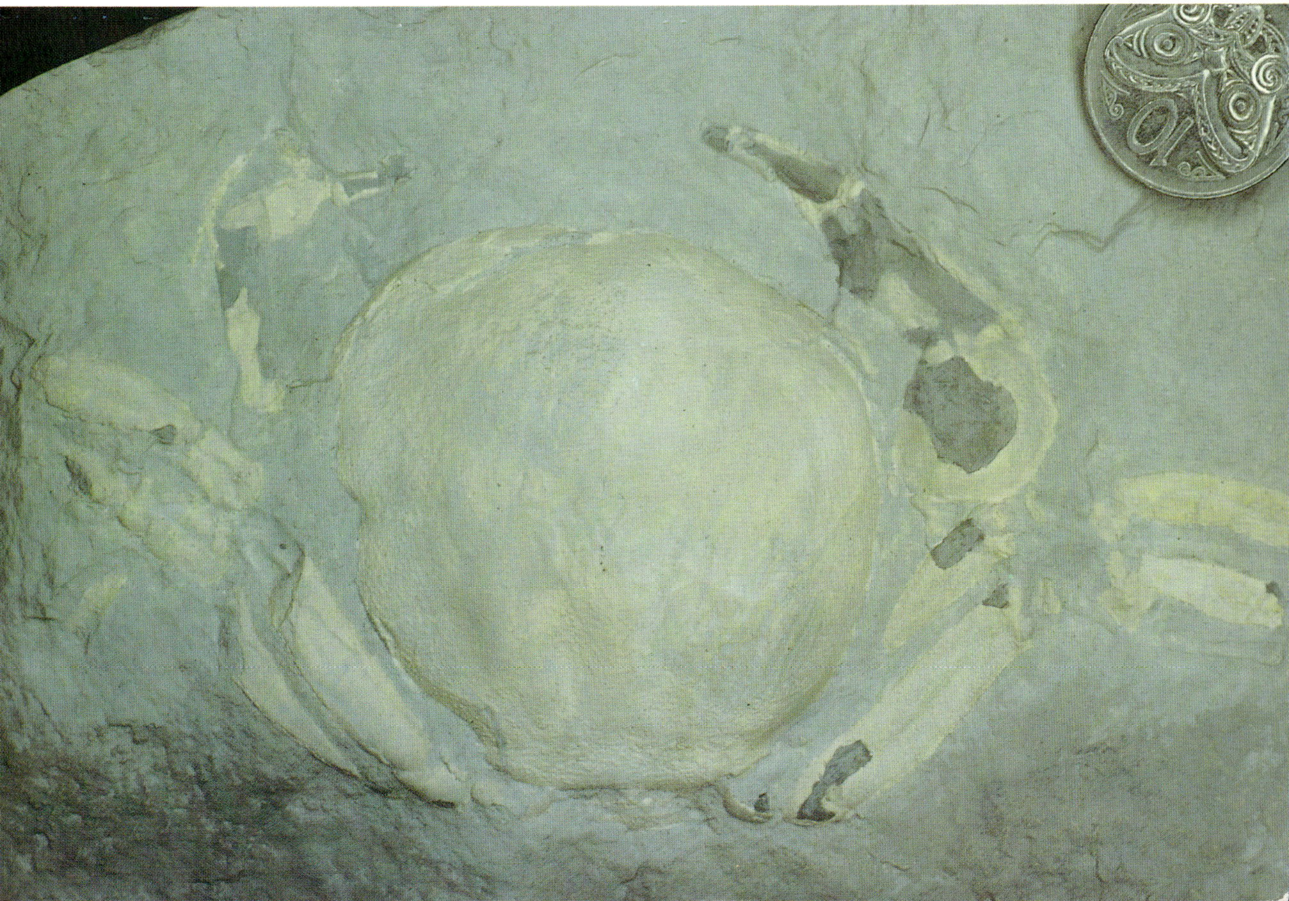

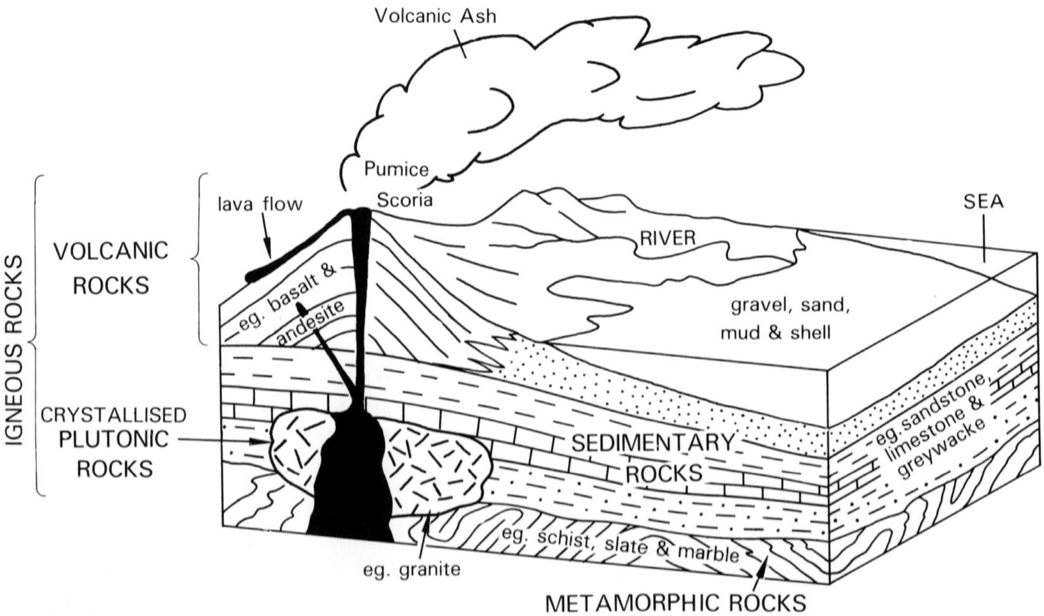

8. Schematic diagram showing how the three major kinds of rocks form.

Millions of dead plants and animals are buried in these accumulating sediments and a few encounter favourable conditions for their preservation as fossils. Several special kinds of sedimentary rock, such as coal and limestone, are formed almost entirely by the accumulation of fossil material. Fossils are not always easy to see in them because of changes these rocks have undergone since burial.

When sediment is first deposited it is usually soft and porous. As it is buried by more and more layers, this sediment is transformed into rock by compaction into a denser and harder material. At the same time minerals carried in solution in water passing through the sediments may be deposited in the pores and contribute to the hardening into rock.

Fossil-bearing sedimentary rocks occur throughout most of New Zealand, although they are uncommon in the Bay of Plenty, Taupo, Wellington, Marlborough Sounds, Southern Alps, Fiordland and Stewart Island areas.

Igneous rocks

Igneous rocks are formed by the cooling and solidifying of hot molten magma. This magma is generated deep within the earth and rises towards the surface up zones of weakness through preexisting rocks. Sometimes the magma solidifies underground before reaching the surface and forms coarsely crystalline plutonic rocks, such as granite. These rocks never contain fossils. They are widespread in Nelson, Westland, Fiordland and Stewart Island.

Magma is a mixture of liquid rock, crystals and dissolved gas. If it reaches the surface, it is erupted from a volcano and cools rapidly to form fine-grained volcanic rocks. Magma with low gas content erupts as molten lava which flows like hot treacle down the volcano slopes and across the countryside before cooling and solidifying into a hard, dense rock, such as basalt or andesite.

Sometimes magma may be full of dissolved gases which, on reaching the surface, are released explosively, blowing the lava into fragments that are ejected into the air as ash, pumice or scoria. This ejected material forms layers (called tuff) that blanket the ground, usually with thick deposits nearby and a thinner covering further afield.

Rocks formed by volcanic eruptions (ash showers, lava flows) only occasionally contain fossils within them. More commonly fossilised vegetation may be preserved buried beneath the ash or lava.

Volcanic rocks are common in the northern North Island, with smaller areas scattered throughout the rest of the country.

Metamorphic rocks

Metamorphic rocks, such as schist, slate and marble, are sedimentary or igneous types that have been greatly changed by enormous pressures and heat, usually as a result of deep burial or contact with hot magma. These changes often recrystallise and distort most of the rock and may destroy any fossils that are present. Thus the schists that form most of Central Otago, the Southern Alps and some of the Marlborough Sounds lack fossils, but fossils can still be seen in some of the slates and marbles of northwest Nelson or the weakly metamorphosed greywackes of New Zealand's axial ranges.

LEFT TOP
9. Igneous rocks are formed by cooling of molten magma. When magma erupts at the surface it cools to form volcanic rocks such as the lava flows, lahar deposits and ash layers that comprise Mt Ngauruhoe and the surrounding area. Fossil trees are occasionally found in volcanic rocks.

MID LEFT
10. Metamorphic rocks are those that have been greatly modified by enormous pressures and temperatures, usually deep within the earth. Many metamorphic rocks, such as this schist in Central Otago, lack fossils, as those that may have been present have been destroyed during metamorphism.

BOTTOM LEFT
11. Most sedimentary rocks can be recognised by their distinctive layering, as in these sandstone and mudstone beds near Auckland. Most fossils are found in sedimentary rocks, often seen in man-made rail and road cuttings.

TOP RIGHT
12. Layered sedimentary rocks that form sea cliffs, such as these at Cape Kidnappers, Hawkes Bay, are often the most rewarding places for finding fossils. In most, the layers of sand, mud or gravel accumulated on the floor of the sea and in favourable conditions preserve seashells or the remains of other marine organisms.

BOTTOM RIGHT
13. Fossil seashells occur on the very top of 380m high Te Mata Peak behind Havelock North. They provide graphic evidence of the tremendous forces deep within the earth that have pushed these rocks up out of the sea over the last few million years.

Mountain top fossils

People are often puzzled by the occurrence high in the mountains of fossils of marine creatures. These fossils and the rocks in which they occur have been uplifted, often many thousands of metres by forces deep within the earth. These forces are related to New Zealand's position on the boundary between two of the earth's major plates. The 5-35km thick outer layer of the earth, known as the crust, is broken into a jig-saw puzzle of moving segments called plates. New Zealand's very existence and its shape are a result of our position for the last 30 million years astride the collision zone between the Indian-Australian and Pacific plates.

Tremendous forces concentrated along this zone have pushed up high mountain chains, pulled other areas down deeply into the earth's crust and split the country into many parts along major fault lines. The originally horizontal-layered sedimentary rocks have often been broken up into blocks, tilted or crumpled into vast folds by these forces. In places, millions of years of uplift have raised formerly deeply buried rocks, such as the schists and granites. As they rose, erosion has removed the overlying rocks and exposed those beneath.

These major earth movements have been going on for hundreds of millions of years, so that today New Zealand is composed of a mosaic of different rock types, young and old; soft and hard; igneous, sedimentary and metamorphic. You are just as likely to find fossils on mountaintops as you are in seacliffs.

How fossils are preserved

14. Fossilisation normally only preserves the hard parts of an organism, such as the 5cm shell of this 30 million year old (Oligocene) scallop, *Serripecten*, from Kakanui, North Otago. The hard parts of aquatic organisms (live in water) have a better chance of early burial by sediment and subsequent fossilisation, than do those of land dwellers.

When an animal or plant dies its softer parts decay rapidly in contact with air and bacteria. Harder parts usually break down over a longer period.

Fossilisation requires special conditions. In most cases only hard parts become fossils. Firstly they need to avoid being broken up into tiny fragments, such as shell hash on a beach, or being eaten by predators. In general early burial is required to protect the fossil from decay, weathering and other damage. Aquatic (water-living) organisms stand a better chance of early burial and preservation as fossils than land-dwellers (terrestrial). Terrestrial organisms seldom die in a place where their remains are rapidly buried and preserved.

Places where organisms are buried and fossilised:

Sea bed

The majority of New Zealand fossils have been preserved by burial in sediments on the sea floor. Very few fossils are preserved in beach deposits because the waves rapidly break beach material down into fine fragments. Most are preserved in calmer places, such as estuaries, lagoons, quiet bays and especially in deeper water on the continental shelf and in the deep ocean. The remains of sea-dwelling organisms are quite frequently buried and preserved before being eaten by scavengers. These fossils include organisms that lived within and on the sea-floor sediments, and also free-swimming forms such as fish, and the remains of plankton that rain down from above. Occasionally the remains of land-dwellers wash into the sea and become preserved in marine sediments.

15. Most New Zealand fossils are marine shells that were buried by sediment on the floor of the sea. Very seldom are shells on a beach preserved as fossils, because they are continually being broken up by waves and uncovered by storms.

16. Most fossils of land-dwelling organisms in New Zealand are leaves, branches or logs that were buried by mud, sand or gravel in ponds, swamps, lakes or river plains.

Lakes, swamps and river plains

Most fossils of land-dwelling organisms in New Zealand occur in sedimentary rocks that accumulated on the floor of freshwater ponds, lakes and swamps or in alluvial sediments of river-plains. The anaerobic conditions of swamps and peat bogs exclude air almost immediately and organic remains are usually well preserved in them.

Sand dunes and landslides

The wind-blown sand of sand dunes sometimes advances into scrub and forest, burying and preserving it. Bones of dead birds and reptiles are also sometimes buried in sand dunes. At times landslides too can entomb trees and other organisms in their paths and these may then be fossilised.

Lava flows

Molten lava generally incinerates any vegetation or animals it engulfs and therefore very rarely do lava flows preserve any traces of life within them. Sometimes the underground roots and lower stumps of forest trees or swampy mud life may be baked, carbonised and preserved beneath a solidified lava flow that had flowed in, cooked and buried them.

Ash showers

Ash showers often kill and bury nearby vegetation preserving tree stumps, felled trunks and sometimes even leaves and twigs as fossil forests.

17. Only rarely do lava flows preserve the remains or traces of animals or plants. Fossil moulds of a number of tree trunks can be seen in a lava flow from Pupuke volcano, in the low cliffs north of Takapuna Beach, Auckland. The trees were picked up by the molten lava as it flowed through a forest. The enormous heat incinerated the logs but not before the cooling and congealing lava had preserved their shape as moulds.

18. Logs are commonly found fossilised in volcanic lahar deposits such as these, which form the ring plains around Tongariro and Ruapehu in the central North Island. The logs were growing trees plucked out by the giant slide (lahar) of volcanic rocks, ash and water, as it sped down the side of the volcano.

Volcanic lahars

Giant slides of ash, volcanic rock and water, called lahars, sweep down the slopes of large andesite volcanoes, such as Ruapehu, and come to rest on the surrounding lowlands forming a ring plain. Like ash showers, lahars may also kill, bury and fossilise forests or, more commonly, rip them up and carry the logs along in the moving slide. Any animal life that is engulfed by a lahar is also a potential fossil. Lahars that flow into seawater or down the slopes of submarine volcanoes often engulf sea life on the sea floor and carry it along to be buried and possibly fossilised in the lahar deposit when it comes to a stop.

Caves

In the extremely dry conditions of some caves (e.g. lava caves), the remains of organisms (such as bird bones) are sometimes mummified and preserved, but seldom survive more than a few thousand years. In wetter caves, remains may be preserved by burial in sediment deposited by subterranean streams or by a coating of lime deposited over the remains by percolating water in limestone caves.

Mineral springs

Hot water often deposits minerals around thermal springs forming sinter terraces. Plant remains, such as leaves, may be coated by these minerals, preserved and eventually buried and fossilised in the sinter.

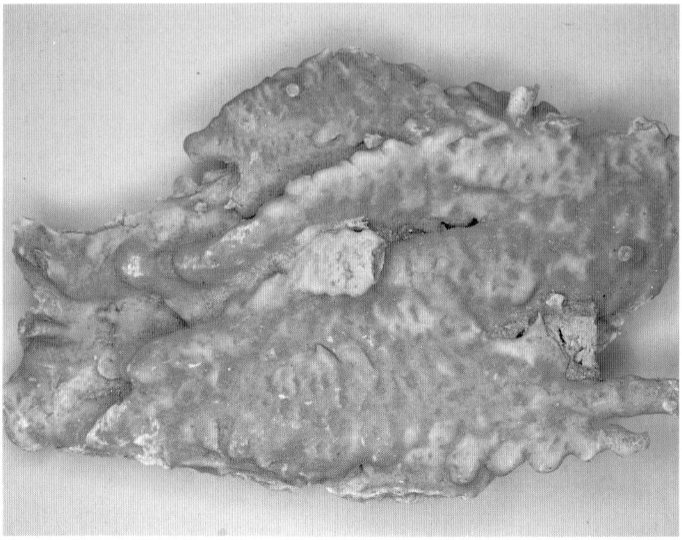

19. In thermal areas and limestone caves, the remains of animals or plants are occasionally coated in minerals and preserved. These fern fronds have been coated in lime by percolating water and fossilised in the entrance of a small limestone cavern. (20cm long)

20. The calcite shells of many barnacles fall apart when the soft parts decay and are usually found fossilised as separate plates. *Bathylasma aucklandica,* 20 million years old (Miocene), Motuihe Island, Auckland.

21. The flattened and carbonised remains of the chitinous skeletons of ancient plankton animals called graptolites, from 450 million year old (Ordovician) rocks in northwest Nelson. (Slab 15cm across)

22. This fish skeleton has been preserved intact in this 20 million year old sandstone from inland Poverty Bay. (12cm long)

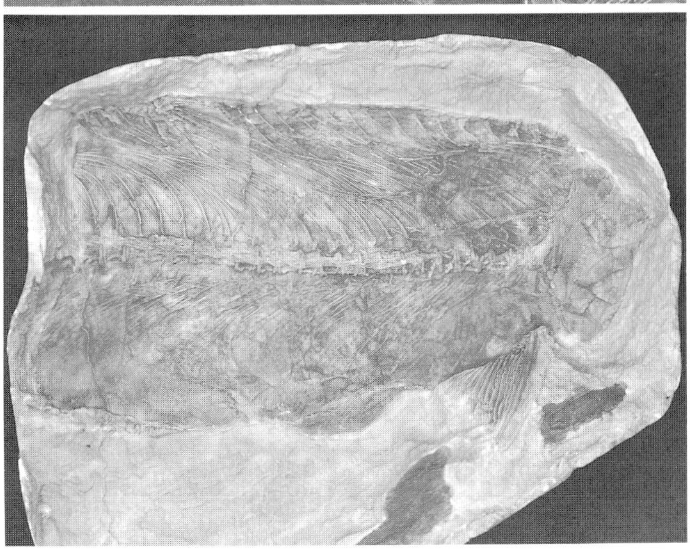

The formation of fossils

There are a variety of ways that a fossil may be preserved.

Mummification

This is the complete preservation, in relatively undamaged condition, of both the hard and soft parts of an organism. Examples are the mummified remains of moas in peat swamps or preservation of complete insects and spiders in kauri gum. The conditions for forming mummifications are rare.

Skeletons and shells

The most common kind of fossil consists of the hard skeletal parts of organisms. These may be composed of bone, wood, chitin, calcite (calcium carbonate) or silica. They include internal skeletal parts (e.g. vertebrate bones, sponge spicules) and external skeletons or shells (e.g. crabs, sea eggs, seashells). These hard parts may be preserved unchanged in their composition, although they usually lose their colour. The hard parts of many organisms (e.g. seashells) are in one piece and are fossilised whole. Others are composed of many pieces held together by soft tissue which decays and the skeleton breaks apart prior to fossilisation.

Petrifaction and replacement

Minerals in solution, percolating through the rock, may add to or replace the original skeletal constituents of the fossils. The process of petrifaction is the gradual impregnation of fossil remains by these secondary minerals, increasing the weight and hardness of the specimen and, as the word suggests, turning it into stone. In some rocks the percolating solutions dissolve away the original skeleton and replace it grain by grain with a secondary mineral. Common secondary minerals are silica (e.g. silicified wood), calcite, iron pyrites, limonite and haematite. In some situations the original mineral composition is unaltered but it is recrystallised to a coarser variety.

23. Calcite (calcium carbonate) shells are the most common kinds of fossils in marine sedimentary rocks, eg. this 5cm lampshell, *Neospirifer wairakiensis,* from Letham Burn, Southland, found in 250 million year old rocks (Permian).

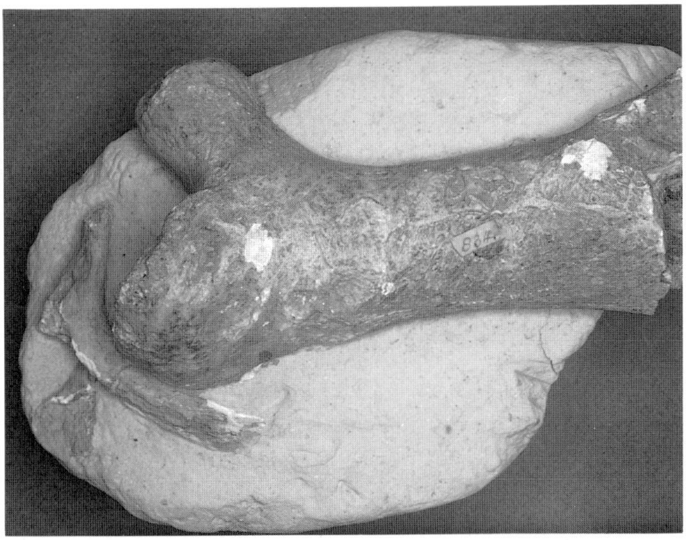

24. The fossilised bones of vertebrates, such as this large, 15cm long, moa leg bone from Scinde Island, Napier (2 million years old), are often found singly. Occasionally complete skeletons are found fossilised and undisturbed since the animal died.

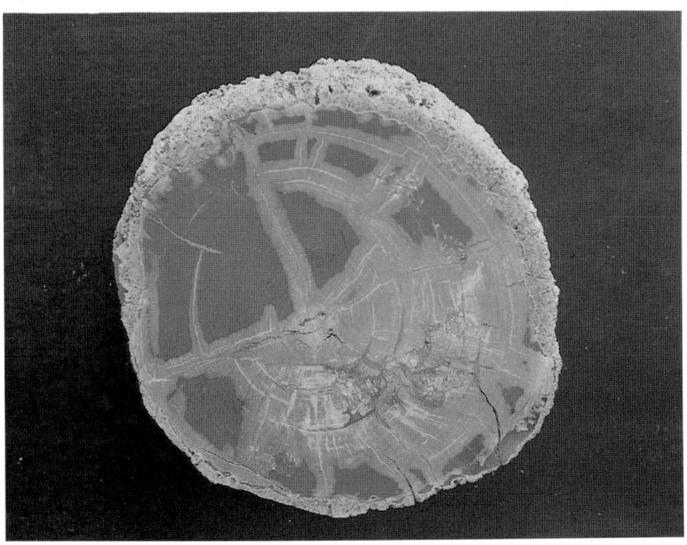

25. The petrified trunk of an 80 million year old (Cretaceous) tree that has been replaced by silica. It has been cut in half with a diamond saw to show some of the growth rings still preserved. (25cm diameter)

26. The impression of a large, 20cm long leaf of the 170 million year old (Jurassic) plant *Taeniopteris crassinervis*, from Mataura Falls, Southland.

27. The black impressions of partly carbonised leaves of the 40 million year old (Eocene) plant *Cinnamomum waikatoensis* from Pukemiro, Waikato. (leaves 10cm long)

Carbonisation

Woody or chitinous material is often fossilised by a process called carbonisation. The wood or chitin is decomposed by anaerobic bacteria resulting in the loss of oxygen and nitrogen and leaving a fossil rich in carbon and usually very dark in colour. Carbonised fossils are usually compressed and flattened during this process. Similar carbonisation changes are involved in the conversion of masses of plant material into peat and coal.

Impressions and traces

Just as footprints may be left in soft mud, the impressions of any organism can be preserved in certain circumstances in mudstone or fine sandstone. These are usually the impressions left by dead organisms (e.g. jellyfish or leaves) that decayed away soon after burial and before the sediment began to harden. Quite frequently the traces (tracks, burrows, borings) left by moving animals, on or in sediments, are also preserved in rocks.

Casts and moulds

These form when the rock containing the fossils has hardened somewhat and subterranean solutions then completely dissolve away the original fossil. This leaves an external mould of the fossil, which may sometimes be found in rocks.

If the fossil was hollow inside (e.g. seashell), it may have filled with sediment during burial. When the hard parts dissolved, an internal fossil cast may have been left.

28. The hard parts of many fossil shells are sometimes dissolved away by water percolating through the rock. This leaves an empty fossil mould of the outside shape of the shell. If the shell had filled with mud or sand when it was buried, then a cast of the inside shape of the shell may also be left, such as this 380 million year old (Devonian) lampshell cast from near Reefton. (6cm across)

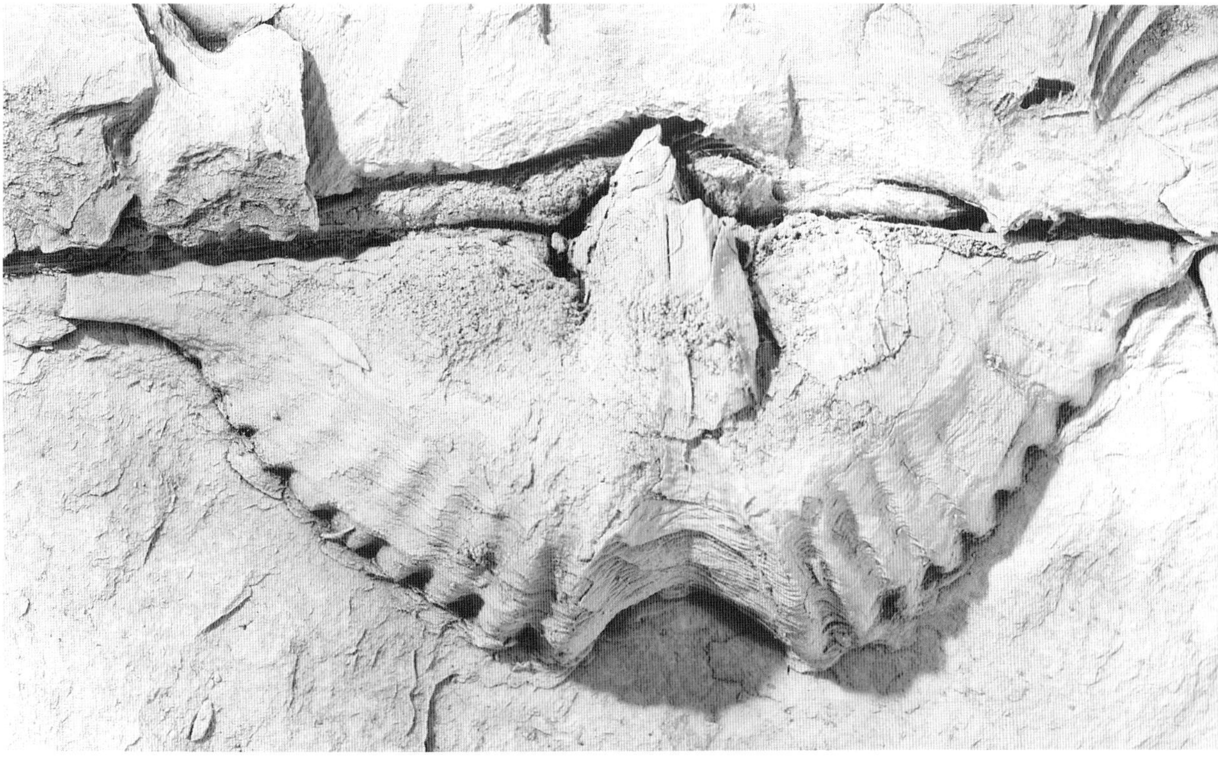

Kinds of fossils

Fossils can be divided conveniently into macroscopic forms, microscopic forms and trace fossils.

Macrofossils

These are fossils that can be seen easily with the unaided eye.

Animal macrofossils are predominantly shells (e.g. bivalves, snails, lampshells), but also include bones (e.g. dinosaurs, whales, moas), teeth (e.g. sharks, fish), chitinous external skeletons (e.g. crabs, trilobites), tubes (e.g. marine worms), calcified external and internal skeletons (e.g. barnacles, corals, belemnites), fish scales, snail operculae (e.g. catseye), and, rarely, eggs and the impressions of soft parts.

Plant macrofossils are mostly the woody branches, trunks, stumps and roots of trees (mummified, petrified or carbonised) and leaves or fronds of trees and ferns. Sometimes seeds, cones or fruit are also found fossilised. Macrofossil remains of lichens, liverworts and fungi have not yet been found in New Zealand rocks although rare moss fossils have.

29. The spiral shells of marine snails are frequent fossils in New Zealand rocks. *Struthiolaria frazeri* from 2 million year old rocks (Pliocene), Whanakino Stm, inland Hawkes Bay. (6cm high)

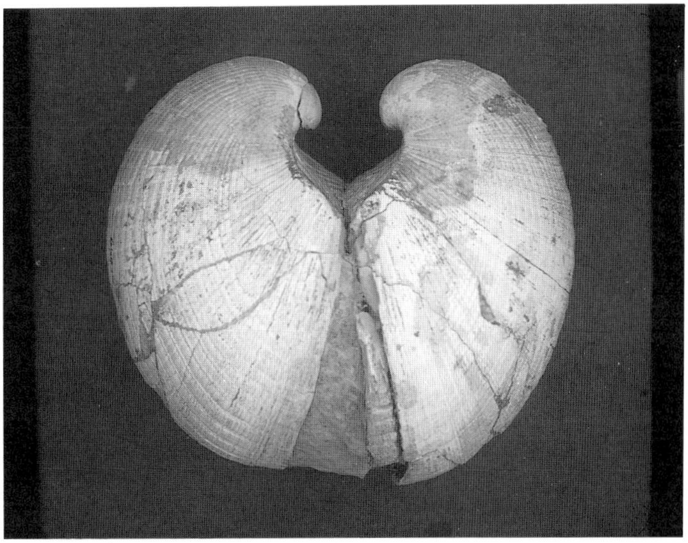

30. Macrofossils are easily seen with the unaided eye and are predominantly shells, like this large 10cm bivalve, *Cucullaea ponderosa* (Karamea, 15 million years old, Miocene) with its two valves still joined together.

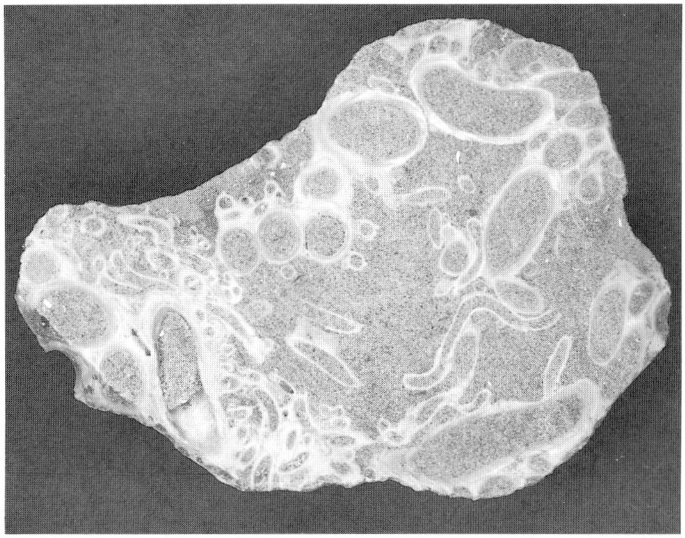

31. This block of 3 million year old (Pliocene) sandstone from Motunau, north Canterbury, has been sawn in half to show numerous calcite tubes of marine worms that have been fossilised within it. (10cm across)

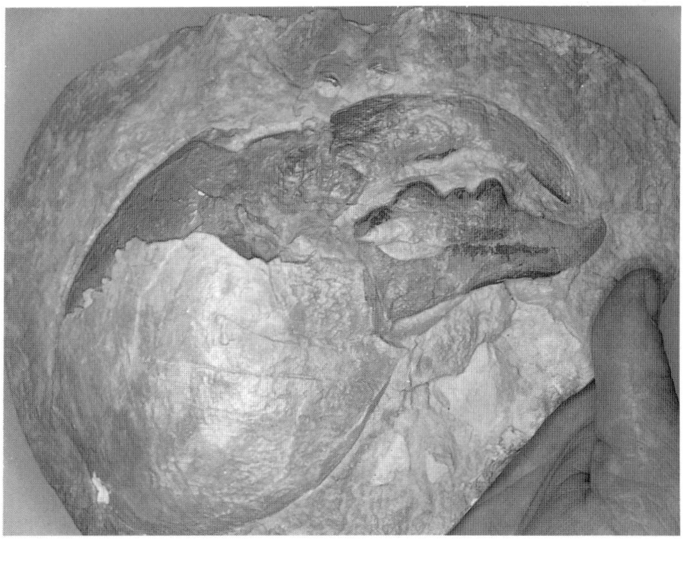

32. This enormous claw of an extinct 15 million year old (Miocene) crab (*Tumidocarcinus giganteus*) is an example of fossilised chitinous hard parts. As the crab grew, one claw expanded rapidly until it was larger than the rest of the animal combined.

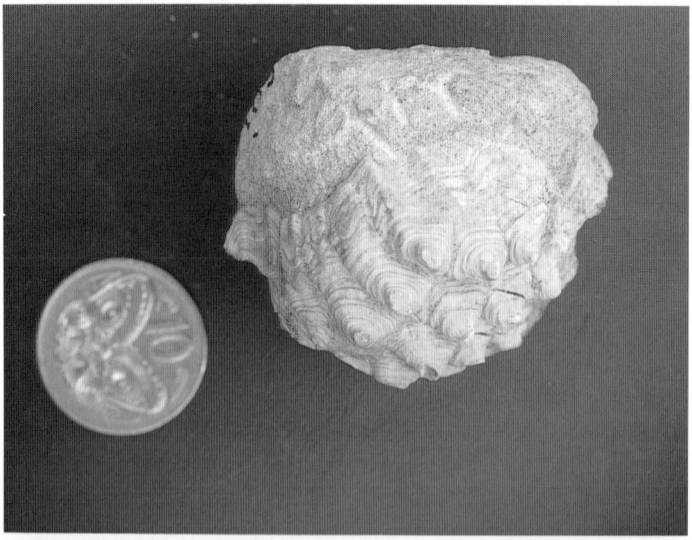

In addition to shells and bones, macrofossils include a wide variety of other forms, such as:

33. Calcareous alga, Claverley Stm, Kaikoura coast, 70 million years old (4cm diam).

34. Fish scale, Kaipara, Northand, 75 million years old (1cm diam).

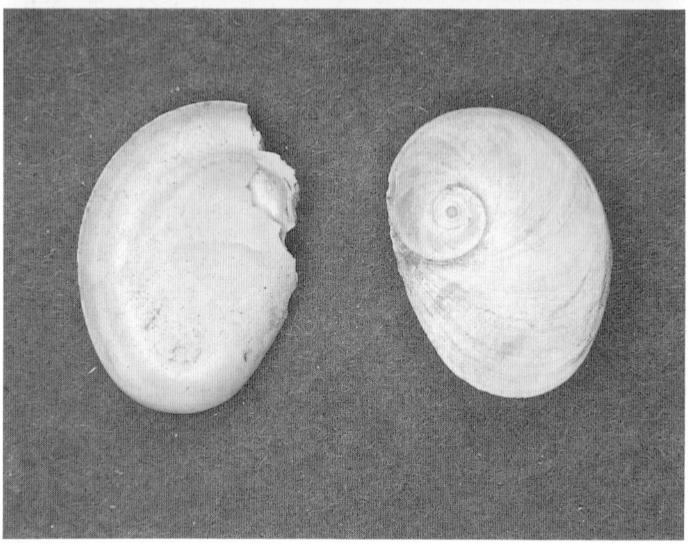

35. Snail opercula, Wendon Valley, Southland, 30 million years old (2cm each).

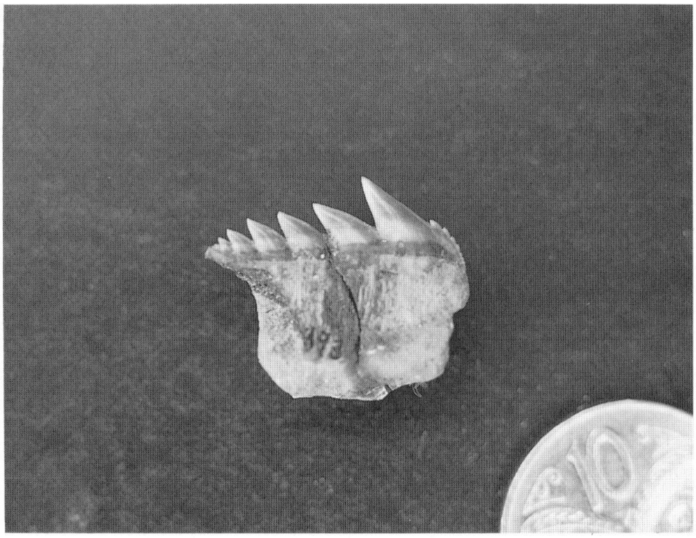

36. Fish teeth, Ohope, Bay of Plenty, 1.5 million years old (1.5cm across).

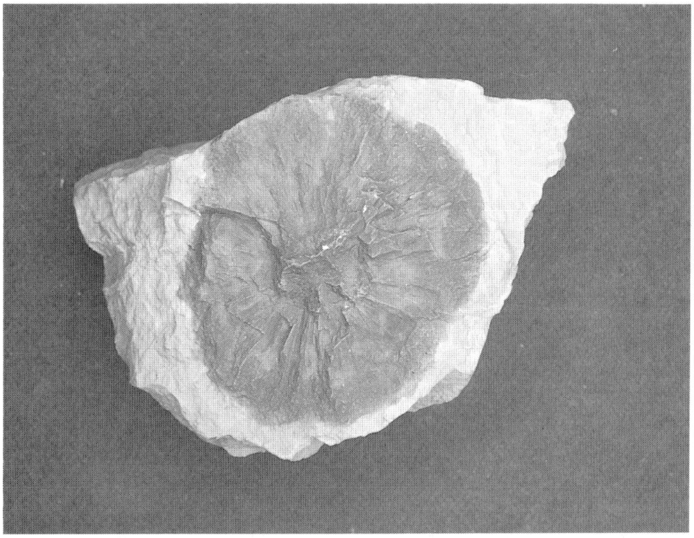

37. Conifer cone, Wairarapa, 75 million years old (4cm diam).

38. Silicified log, Whangaroa, Northland, 20 million years old (1m long).

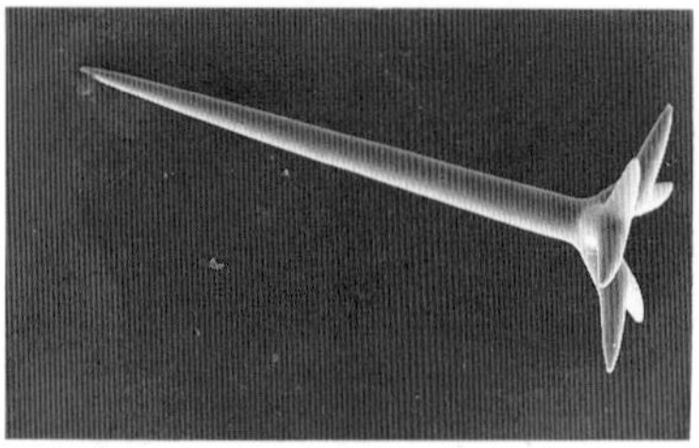

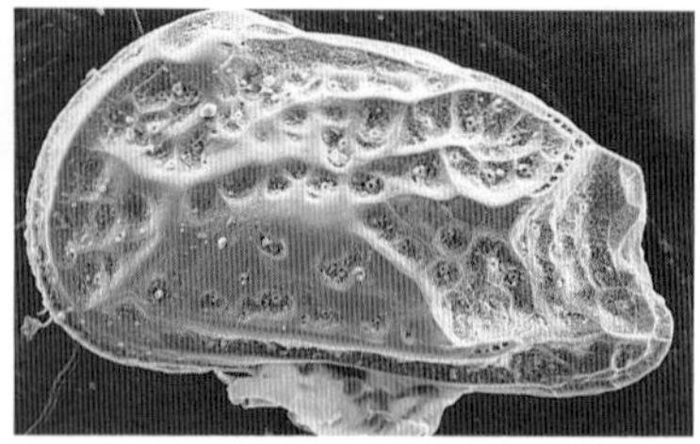

39-41. Top row (left to right): internal spicule of a sponge (0.05mm); otolith or fish ear bone (1mm); ostracod or water flea (0.5mm).

42-44. 2nd row: bottom-dwelling marine protozoa or foraminifera (0.4mm); plankton-dwelling foraminifera (0.3mm); plankton-dwelling protozoa or radiolarian (0.1mm).

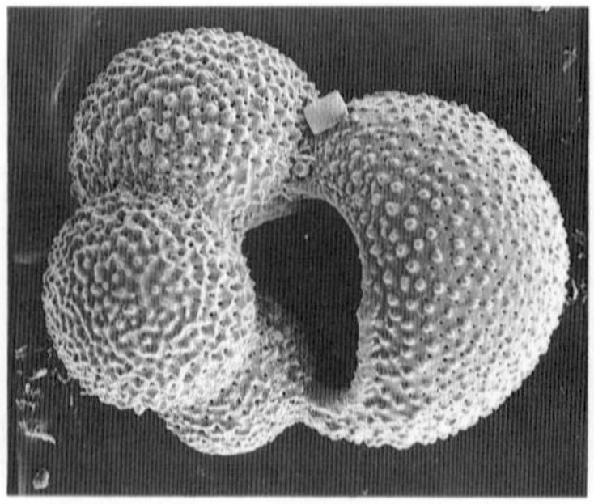

Examples of the wide variety of forms preserved as microfossils in rocks and studied using a microscope:

Microfossils

These are fossils that can only be seen in detail using a microscope. They are generally smaller than about 1mm in size, and many are much smaller than this.

Animal microfossils include the small hard parts of larger organisms, such as small fish teeth, fish ear-bones (called otoliths), worm jaws, spines of sea eggs, hard internal skeletal spicules of sponges, sea cucumbers and starfish or the external skeletons of small animals, such as waterfleas (ostracods).

More common microfossils are the hard parts of some groups of marine protozoa (one-celled organisms), called foraminifera and radiolaria. Some of these live in sediment on the sea floor and many more live in their teaming millions floating around in the plankton of the world's oceans. Their minute skeletons sink to the bottom of the sea when they die and are readily preserved as fossils.

The most useful plant microfossils are the hardy pollen grains of flowering plants and conifers and the spores produced by ferns and fungi. Plants that are wind-pollinated in particular, produce millions of tiny pollen grains each season. They are blown around and

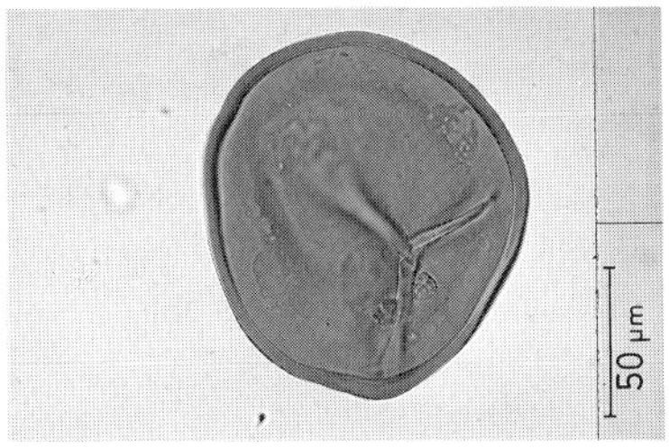

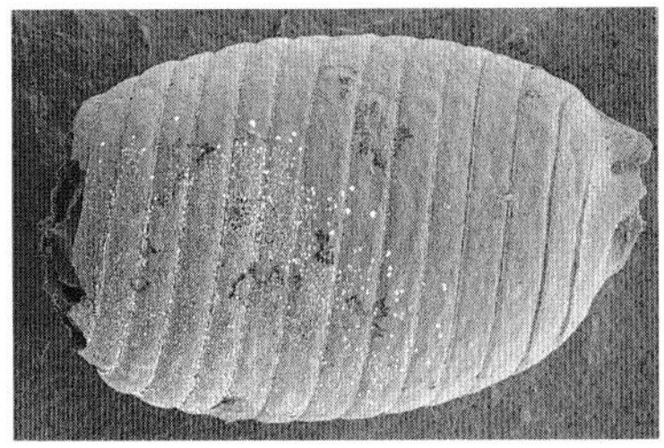

45-47. Top row (left to right): pollen grain of a flowering plant (0.05mm) Photo: Dallas Mildenhall; fern spore (0.05mm) Photo: Ian Raine; reproductive part of a freshwater alga or stonewort (1mm).

48-50. Bottom row: siliceous marine alga or dinoflagellate (0.1mm) Photo: Graeme Wilson; siliceous alga or diatom (0.05mm); calcareous marine alga or coccolith plate (0.005mm) Photo: Tony Edwards.

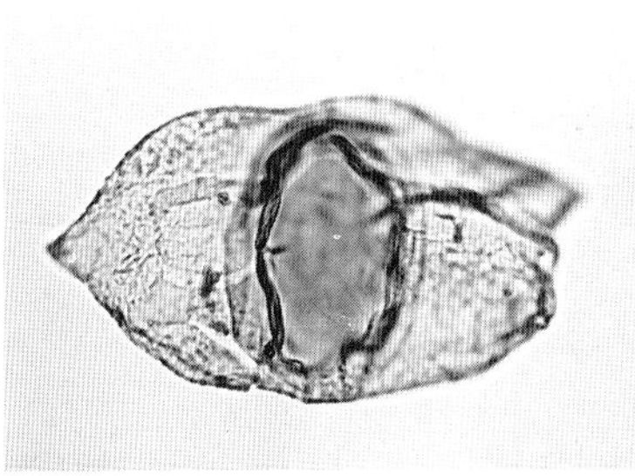
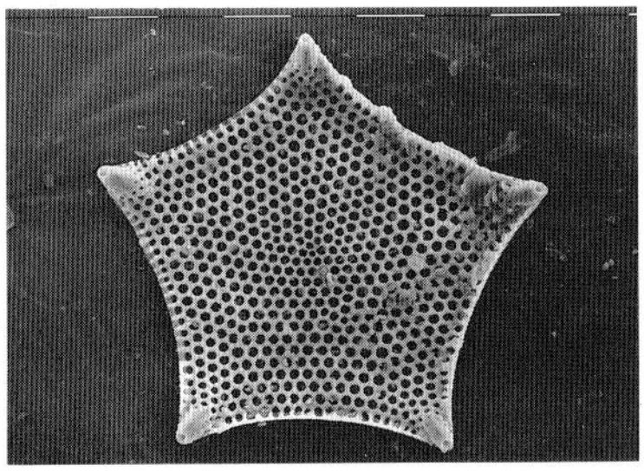

carried by water into sedimentary environments where they are preserved (mummified) in vast numbers.

Several groups of microscopic, single-celled, marine and freshwater algae produce hard skeletons made of silica (e.g. diatoms, dinoflagellates) or calcite (e.g. coccoliths). These are found fossilised in huge numbers in marine and freshwater bottom sediments. Many of these algae are also part of the plankton that float around in the sea or lakes and in certain favourable conditions produce algal blooms.

The remains of microscopic pollen, spores, marine algae and protozoa are numerically by far the most abundant fossils to be found in rocks. A hand-sized specimen of most marine or freshwater sedimentary rocks (e.g. mudstone, sandstone, peat) contains thousands to hundreds of thousands of these small fossils. They are so small that they are seldom seen except by specialist paleontologists who use special preparation techniques and high powered microscopes to study them.

Trace fossils

The traces of moving animals are often preserved in sedimentary rocks. These may be tracks or imprints made on the surface of fine-grained sediment, that have been captured by rapid burial; they may be the back-filled burrows made by active sediment-eaters such as worms, heart urchins and other animals; or the shelter burrows, sometimes with an internal lining, used as retreats by some deposit and filter feeders, such as mud crabs, snapping shrimps and some worms.

All these kinds of trace fossils are best seen in rocks consisting of alternating beds of different colour or grain-size, so that the tracks or burrows are filled with sediment that contrasts with the host rock.

Trace fossils also include the borings in rock, shell, wood and even leaves made by certain bivalves, sponges or insects. Fossilised faecal pellets and coprolites (droppings of animals) are also included in this category, as are crop or gizzard stones. Moa gizzard stones are frequently found in young sand dune, ash shower and peat deposits in New Zealand.

51. Fossilised footprint, probably of a kiwi, preserved in 20 million year old rocks (Miocene), north Westland.

52. Polished and rounded, white quartz pebbles (0.5-6cm in size) found in soil, dune sand and young rocks, are often the gizzard stones of moas that roamed New Zealand thousands or hundreds of thousands of years ago.

53. Fossil burrows of sediment-eating worms in Miocene mudstone near Auckland. These trace fossils stand out because they are filled with sand from the overlying layer. (Photo is 50cm across)

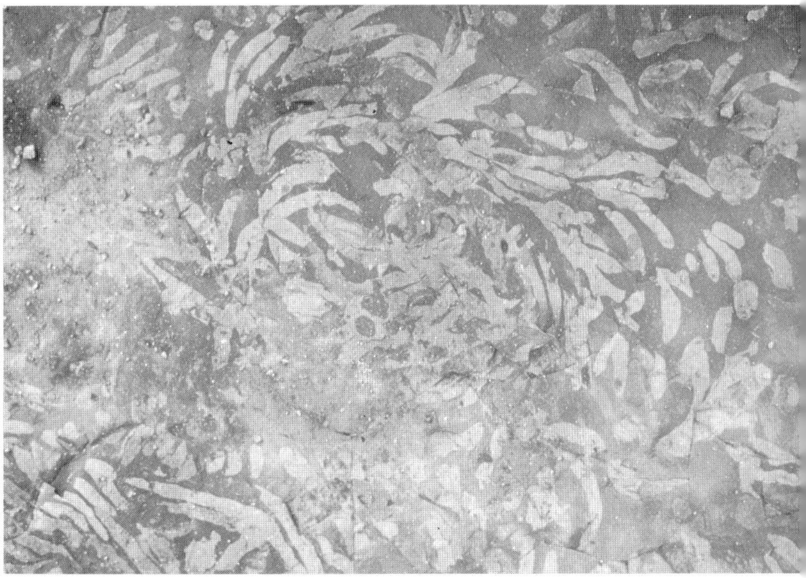

54. Fossil coprolite (animal dropping) from Jurassic rocks. (length 4cm)

55. A slab of hard, 80 million year old (Cretaceous) sandstone from Mikonui Stream, Kaikoura, containing numerous fossil belemnites (*Dimitobelus lindsayi*) — the internal skeletal parts of extinct squid.

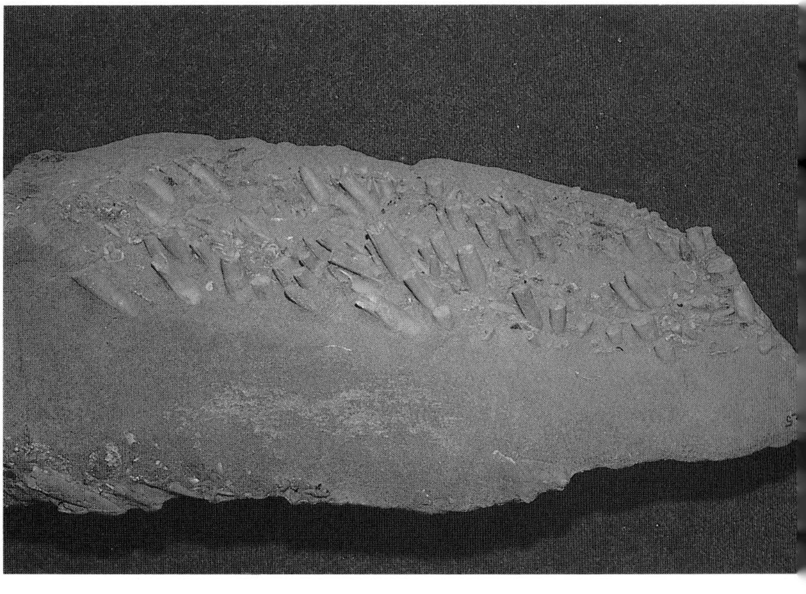

Classification of fossils

Fossils are classified into groups the same way as present life forms, although some additional groups of extinct animals and plants are also recognised. All life forms are placed in one of five kingdoms – the animal kingdom, plant kingdom, fungi kingdom, bacteria kingdom and a group containing unicellular organisms intermediate between plants and animals, called protists. Each kingdom is divided into phyla (animals, protists) or divisions (plants, fungi, bacteria).

The main phyla and divisions with fossil remains in New Zealand are:

Kingdom Animalia

Porifera	— sponges
Coelenterata	— jellyfish, corals
Bryozoa	— moss animals
Brachiopoda	— lampshells
Mollusca	— bivalves, snails, squid, chitons, tusk shells, ammonites, belemnites
Annelida	— worms
Arthropoda	— trilobites, crabs, barnacles, insects
Echinodermata	— sea eggs, starfish, sea lilies
Hemichordata	— graptolites
Chordata	— fish, amphibians, reptiles, birds, mammals

Kingdom Plantae

Thallophyta	— algae, fungi, lichens
Bryophyta	— mosses, liverworts
Pteridophyta	— ferns, horsetails, lycopsids
Spermophyta	— conifers, angiosperms (flowering plants)

Kingdom Protista

Cyanophyta	— blue-green algae
Protozoa	— foraminifera, radiolaria

Fossil names

Every kind of organism has been given a Latinised scientific name consisting of two parts - the genus name (with initial capital) followed by the species name (lower case). Thus human beings are known as *Homo sapiens* (*Homo* = genus name, *sapiens* = species name).

56. Sedimentary and volcanic rocks usually accumulate in layers one upon the other, such as in these 30m high sea cliffs at Waiwiri Beach, south of Port Waikato. The lowest layers contain the oldest fossils (25 million years old) and they get progressively younger upwards with the youngest fossils (20 million years) in the topmost layers.

What fossils tell us

Fossils record the history of life on earth and are the best evidence of evolution and how it proceeded. They provide the most widely used method for determining the age of rocks and also give valuable information about past environments, which can be used to reconstruct a picture of the changing geography of various regions. In addition fossils are widely used in the study of past climates.

The history of life and evolution

Rocks are generally laid down in layers, one upon the other, on land, in lakes or on the sea floor. Unless disturbed by earth movements, the oldest rocks are on the bottom and the youngest on top. It follows therefore that the oldest fossils are preserved in the lowest layer and the youngest are found in the top layer, with a whole succession of fossils of intermediate ages in between.

Nowhere in the world is it possible to find a complete, unbroken succession of rock strata that spans all of earth's history, although in some places like the Grand Canyon (USA), a vast period of time is represented. In most places in New Zealand the rock successions that can be studied in seacliffs, river beds or mountain bluffs, were deposited during only small portions of geological time (a few million or tens of millions of years). The fossils in hundreds of these short sections have been studied and when the results are put together they all show the same consistent succession of fossils through time. This succession forms the basis of the geological time scale and the record of evolution.

The value of fossils for studying evolution is often downgraded by biologists, for they point out that the fossil record is incomplete and patchy. This is true, particularly for soft-bodied organisms and land-dwelling animals, which have only a small chance of being buried and fossilised. Despite this, in many of these groups there are sufficient fossils known to give a broad outline of their evolutionary history, even if intermediate linking forms have not yet been found.

In many marine groups of animals, such as shellfish (molluscs) and especially microfossils (foraminifera, coccoliths, radiolaria etc), the fossil record of their successsion and evolution through time is extremely good with very few gaps. As more and more fossils are found and studied in detail our knowledge of the succession of life on earth continues to improve.

Over long periods of geological time, organisms changed in character, others branched out into new forms and many large groups died out altogether. Many of these changes are described in some detail later in this book, in sections on the history of life in New Zealand.

Age of the rocks

The succession of fossil forms and their record in the rocks has been used to divide up geological time into a time scale or calendar. The international geological time scale is divided into eras and periods and was established using the succession of major fossil groups (phyla, classes) found in British and European rocks. This succession of major groups from trilobites and graptolites through to flowering plants, mammals and finally humans, is the same the world over and allows us to recognise the international eras and periods, even in New Zealand. For example, rocks having the same types of fossils, such as dinosaur bones, are considered to be of similar age and said to belong to the Age of Dinosaurs or Mesozoic Era. Use the accompanying chart as a key to the time scales referred to through this book.

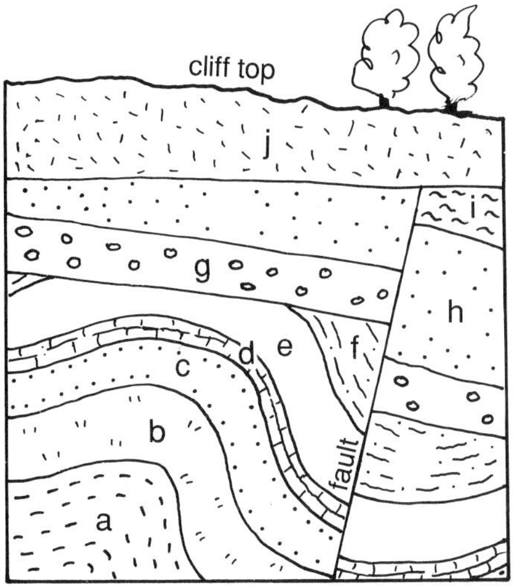

57. Nowhere in the world is there preserved a complete, undisturbed succession of strata and fossils spanning all of geological time. There are, however, thousands of sequences through various parts of time, like this schematic cliff sequence. When studies on the fossils from all these sections are put together they record the evolutionary succession of life on earth. In this diagram the oldest fossils occur in the lowest layer (a) and the youngest fossils are in the uppermost layer (j). Even though the rocks may have been folded, faulted and eroded, with pieces of time not represented by preserved rock layers, it is still possible to determine the sequence of events and the order of deposition of the layers.

58. The International Geological Time Scale. Earth's geological history has been divided into major units called Eras, that have been further subdivided into Periods. These Eras and Periods are recognised in rocks worldwide based on the characteristic fossil groups they contain.

millions of years ago	ERA	PERIOD
0	QUATERNARY	Holocene
		Pleistocene
1.6	TERTIARY	Pliocene
5		Miocene
24		Oligocene
36		Eocene
58		Paleocene
65	MESOZOIC	Cretaceous
140		Jurassic
190		Triassic
235	PALEOZOIC	Permian
300		Carboniferous
345		Devonian
390		Silurian
430		Ordovician
500		Cambrian
570	Precambrian	
4500		

Detailed studies of the earth's history, or the rocks of even a small area, require a much finer subdivision of geological time. This has been achieved by close study of regional successions of fossil types, which differ quite significantly between regions. Thus in New Zealand, Mesozoic, Tertiary and Quaternary time (last 235 million years) has been subdivided into 46 regional stages, which are used in determining the ages of our rocks.

The best fossils to use for fine subdivisions of time are those which occur commonly and evolved rapidly into many, short-lived, distinct types. The finely subdivided New Zealand time scale is largely based on fossil marine shellfish. This is because the strata laid down in the sea are less exposed to erosion and provide more complete successions for study. Most of these shellfish are endemic, having evolved in the New Zealand region and not spread to other parts of the world.

There are, however, some groups of fossils that lived a free-swimming or drifting life in the oceans. These occur as fossils worldwide, although not always in great abundance. Some of them (e.g. graptolites, ammonites, foraminifera) evolved rapidly through many different types and are of great value for detailed recognition and matching of rocks of the same age right around the world.

The greatest use of fossils today is in determining the age of rocks. This information is essential for almost all other geological studies, such as geological mapping, compilation of a region's geological history and changing geography, and in the search for hydrocarbons and other resources.

Actual age of the rocks

The geological time scale based on fossils is a relative time scale, in that one can determine whether a rock is older or younger than another, but fossils do not directly give the age in years. The geological time scale, however, has been calibrated against a scale in millions of years and this has been gradually improved using a number of techniques.

The most widely used techniques for determining the actual age of rocks involve minute radioactive clocks preserved within them. These clocks consist of unstable forms (isotopes) of certain elements, such as uranium or thorium, which gradually break down into more stable elements, in these cases lead. Since the rate of breakdown can be calculated it is possible, by measuring the amount of lead produced at the expense of uranium or thorium in the rocks, to determine their age.

Determining past environments

Fossils are commonly used for determining the past environment in which they were buried. In many instances the present is used as the key to the past. Associations of organisms that live together today in particular environments are assumed to have done so in the past.

Fossils are particularly useful in distinguishing the three main environments where rocks are laid down – on land, in fresh water or in the sea. Each of these is subdivided into more precise environments that fossils help distinguish, such as lakes, rivers, ponds and swamps. Land environments are largely dependent on topography and climate, which produce vastly different plant communities and associated animals. Plant fossils, especially their abundant pollen and spores, play a major role in determining past land environments.

In the sea, different environments are largely related to depth, although in shallow water, salinity and the degree of exposure to waves and storms is also important. Fossil bivalves, snails, corals and foraminifera are the most important groups for determining different marine environments of the past.

Fossil associations usually contain the remains of organisms that lived in or close to the environment in which they were buried. Many also contain additional fossil remains of organisms that did not live in that environment and had been transported in after they died. For example rivers carry logs and land animals into the sea, but seldom are the remains of marine organisms carried inland into land or freshwater environments. Many shallow sea-dwelling organisms are washed down into deeper water but rarely are deeper sea-bottom dwellers carried up into shallow environments. Recognition of fossil associations of mixed origins often provide further clues about the nature of the environment in which they were fossilised.

Determining past geography

When the environmental clues given by fossils are put together for a region, or for all of New Zealand, it is possible to reconstruct the geography for different times in the past. The position of the ancient coastline can be determined and areas of mountains, lowlands, shallow sea and deep sea basins can all be plotted. When this is done for every period of geological time, we develop a picture of the changing geography of New Zealand throughout its history.

Determining past climates

Like environments, past climates are also determined using the present as a key — associations of organisms that live today in certain climatic conditions are assumed to have done so in the past. Thus for example, the presence of fossil reef corals, small coconuts and numerous fossil shells of tropical snails in 25-12 million years old (Miocene) rocks in northern New Zealand, provides good evidence of a tropical climate and the warmest period in New Zealand for at least the last 50 million years.

Microscopic pollen and spore fossils from young sedimentary rocks (last 2 million years) are being used increasingly to document the alternating periods of cold glacial and warm interglacial conditions. Each cycle takes about 100,000 years. Studies of subfossil pollen and spores in very young sediment record the arrival and impact of humans in New Zealand over the last 1000 years. A major decrease in the abundance of forest tree pollen, and an increase in bracken and scrub spores and pollen, accompanied the earliest wholesale firing and clearing of the land for cultivation about 600 years ago.

59. Lumps of reef coral, such as this piece from Hicks Bay, East Cape, are found fossilised in 22-16 million year old (early Miocene) rocks near East Cape and in various parts of Northland. By comparison with modern coral reefs which only grow in tropical areas, they show that the sea around northern New Zealand was 5 - 7 degrees warmer then than it is today.

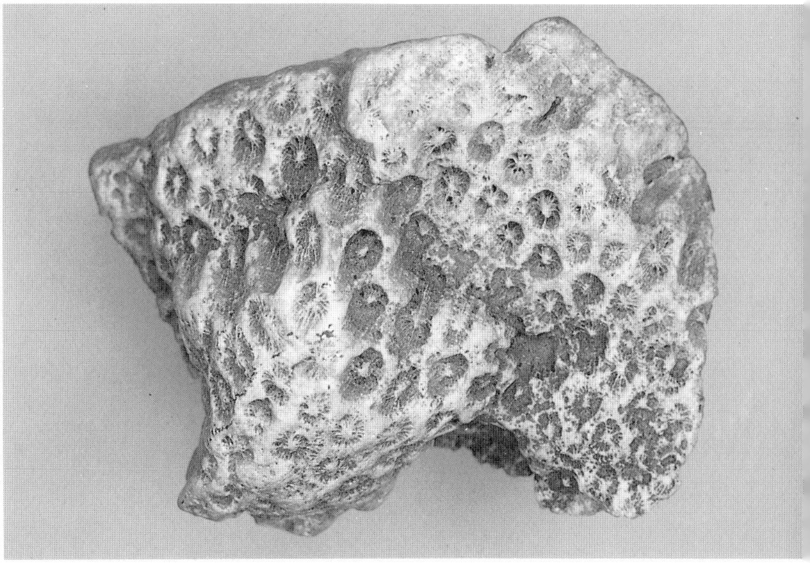

60. Different organisms live in different parts of the sea and their fossilised remains in sedimentary rocks are widely used to determine the salinity, amount of wave action and depth of the environment in which the rock accumulated.

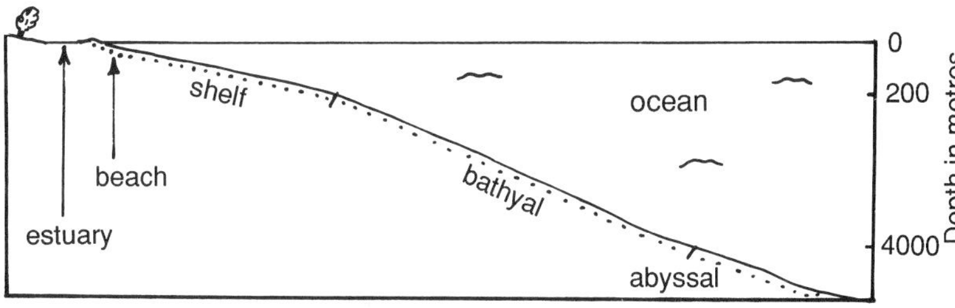

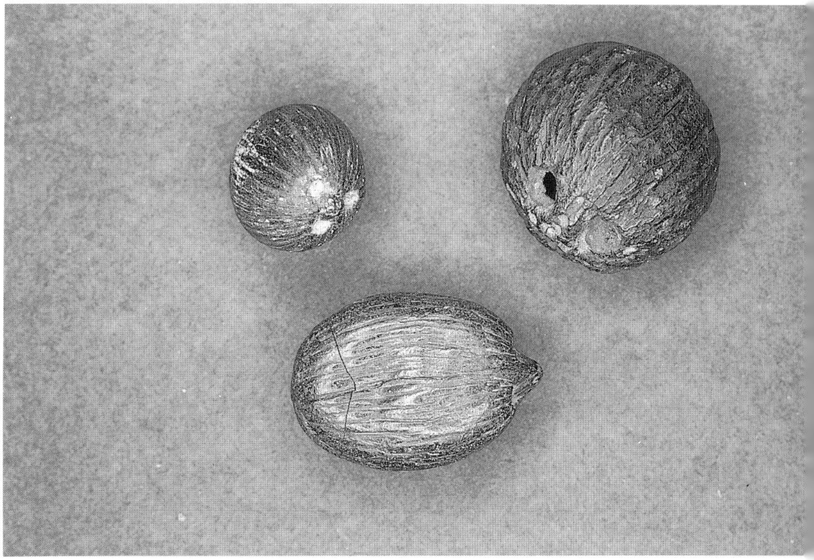

61. Small fossil coconuts erode out of 12 million year old (Miocene) sandstone and lignite and are washed up on Coopers Beach, Northland. They show that the climate was considerably warmer than at present.

Precambrian life – the oldest rocks and fossils

The oldest rocks and fossils on earth are of Precambrian age, but none are known for certain in New Zealand. In Westland, several patches of strongly metamorphosed rock that lack fossils, have previously been thought to be this old, but recent studies indicate that they are younger.

Precambrian fossils are known in a few places in the world and record the beginnings of life on earth. The first signs of life are primitive, single-celled blue-green algae and bacteria, found fossilised in rocks that formed around 3000 million years ago. The first nucleus-bearing cells (green algae) appear in rocks about 1000 million years old. These led to the evolution of multi-celled plants and animals, as represented by the diverse soft-bodied fossils (impressions of jellyfish, worms, seapens) of late Precambrian age.

The evolution of hard skeletons around 570 million years ago marks the end of the Precambrian era and the dawn of a rapidly increasing fossil record.

62. New Zealand's oldest fossils (530 million years old) occur in Trilobite Rock, Cobb Valley, northwest Nelson. This is the easiest place in the country to see fossil trilobites. They are best seen in slightly weathered broken surfaces, rather than freshly broken faces. Like these tail portions of the trilobite *Koptura*, most trilobite fossils in New Zealand are broken and about 3 - 25mm in size. Photo: Roger Cooper. (10cm across)

New Zealand's record of Paleozoic life

The Paleozoic era consists of six periods and spans the time between the appearance of the first skeletal fossils 570 million years ago and a major episode of extinction that killed off well over half the forms of life on earth 235 million years ago. The record of Paleozoic life preserved in New Zealand rocks is rather sparse and incomplete compared with rich fossil assemblages known overseas. This is mainly because Paleozoic rocks are only found at the surface in a few places in New Zealand and most have been greatly hardened, squashed or metamorphosed.

The three oldest Paleozoic periods – Cambrian, Ordovician and Silurian – are sometimes refered to as the Age of Invertebrates, as vertebrates only evolved towards the end of this time. In New Zealand, rocks formed during these periods occur only in the western half of the South Island.

New Zealand's oldest fossils occur in Cambrian limestones of the Cobb Valley and Springs Junction, west Nelson. Most of these fossils are very small and not easily seen in the rocks. Many have been extracted by slowly dissolving away the rock in acid. This ancient fauna contains a wide variety of forms that had rapidly evolved during the early part of the Cambrian. They are all shallow water, bottom-dwelling sea creatures, such as trilobites, sponges, lampshells, conodonts, snails and bivalves.

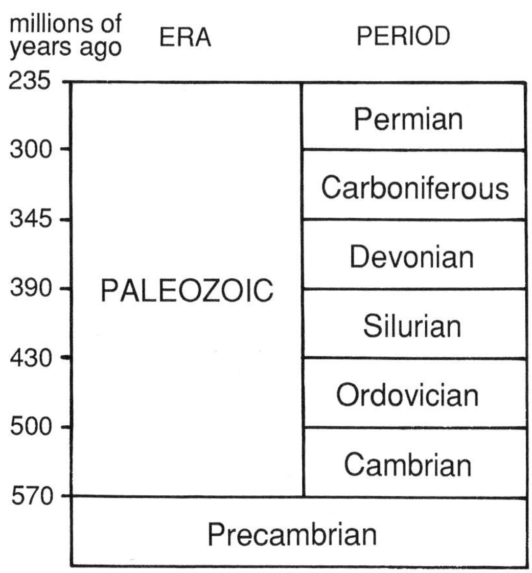

63. The Paleozoic Era is divided into six Periods.

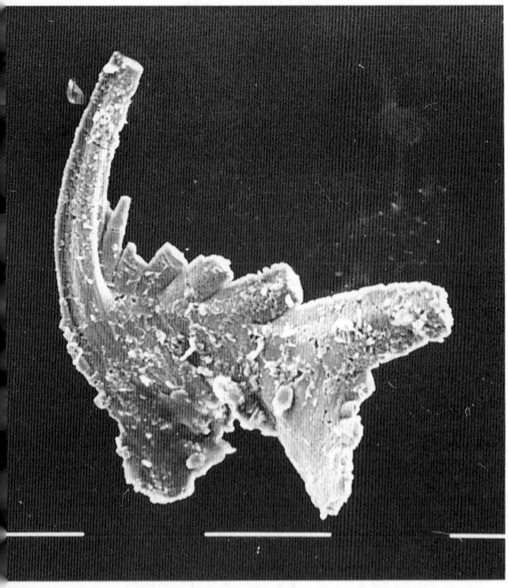

CONODONTS
64. Conodonts, such as this, are a group of microfossils (mostly 0.2 - 1mm) that occur commonly in Paleozoic and Triassic rocks. These small toothed structures are made of calcium phosphate and are particularly valuable in dating Paleozoic rocks overseas and in New Zealand. They are the only fossils of Carboniferous age so far known from New Zealand. Conodonts have been shown only recently to be minute teeth of small extinct sucker eels or lampreys. This conodont is from the Ordovician of northwest Nelson and is 0.5mm long. Photo: John Simes.

Ordovician fossils in New Zealand are dominated by the ancient plankton group called graptolites that are frequently found on the split surfaces of slates, especially in the Cobb Valley but also in other parts of west Nelson and at Cape Providence in Fiordland. Other fossils found in Ordovician rocks here include occasional lampshells, trilobites, conodonts, bivalves and some of the oldest known solitary corals.

The Silurian period witnessed the appearance of the first vertebrate life, the first reef corals and the first terrestrial plant life. In New Zealand the only known Silurian fossils are a small assemblage of lampshells from Hailes Knob in northwest Nelson.

The Devonian period was the Age of Fishes during which a great variety of often armour-plated fishes evolved and dominated the sea and freshwater realms of the world. Scales and plates of some of these primitive fish have recently been found in Devonian rocks near Reefton. As well as Reefton, Devonian fossils are also present in the Baton River, northwest Nelson. In both places most fossil skeletal material has been dissolved away leaving well preserved casts and moulds mostly of lampshells but also corals, bryozoa, sea lilies, snails, bivalves, trilobites and sponges. Towards the end of this period many fish died out and some began to evolve legs and move from water onto land as the first amphibians.

The last two Paleozoic periods, Carboniferous and Permian, are often called the Age of Amphibians, for these animals evolved into many forms and were the dominant land creatures of the time. In Carboniferous seas, sharks had their greatest development and on land a wide variety of fern-like plants was growing. The only known Carboniferous fossils found in New Zealand are microscopic conodonts from Kakahu in South Canterbury.

Volcanic and sedimentary rocks of Permian age are quite widespread in New Zealand forming large areas of hilly country in Nelson, Marlborough, west Otago and Southland. Consequently Permian fossils are more abundant than those from any of the earlier Paleozoic periods. A great variety of lampshell and bivalve fossils has been found, although most are broken, deformed or dissolved and only preserved as moulds, casts or impressions. These fossils have been used to show that the Permian was a time of alternating warm and cool periods, very much like the more recent Ice Ages. Indeed there is evidence of vast ice sheets in Australia, Africa and South America during the Permian. Other characteristic Permian fossils in

65. The tail of a large Devonian trilobite from northwest Nelson.

66. The mould of a strongly ribbed and winged bivalve (*Actinopteria mackayi*) from Devonian rocks at Reefton. (8cm long)

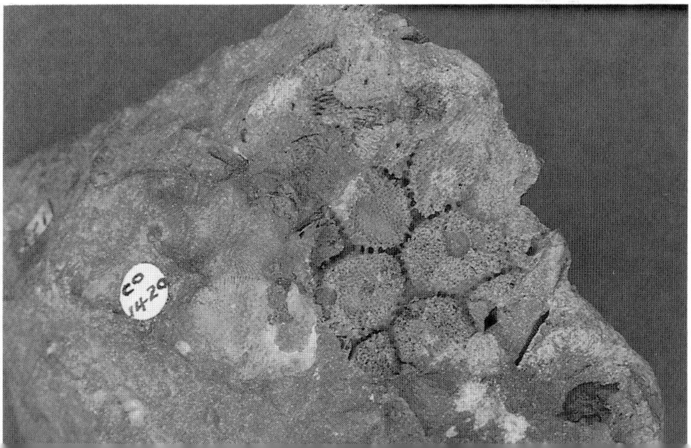

67. The cast and mould of a 2cm piece of colonial coral (*Pleurodictyum megastomum*) from Devonian rocks at Baton River, northwest Nelson.

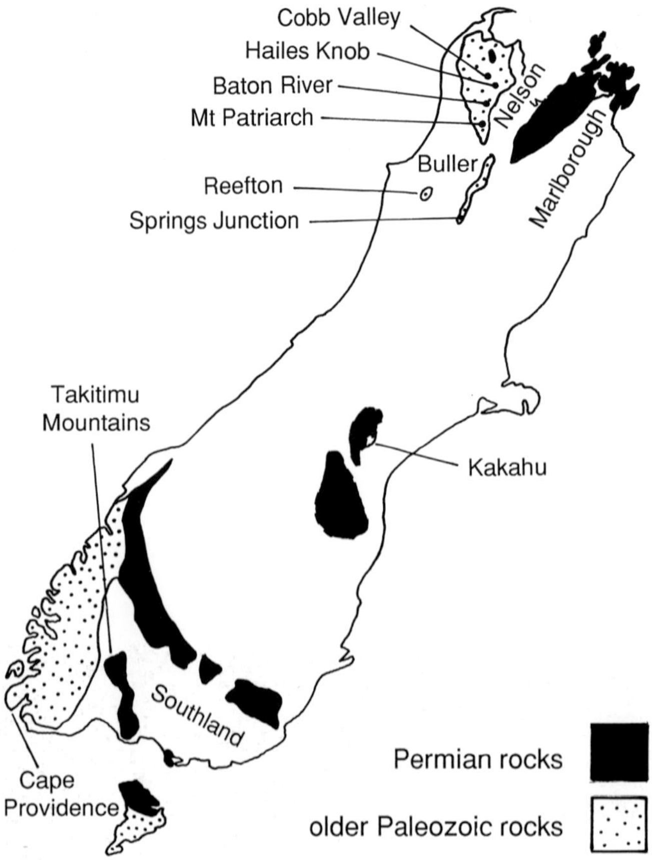

68. Location of Paleozoic rocks in the South Island. The only Paleozoic macrofossils known in the North Island are Permian age and occur in Northland.

New Zealand include sea snails, shelled squids (ammonoids), trilobites, bryozoa and corals.

Fragments of a widely-distributed Southern Hemisphere (Gondwanan) land flora are preserved in Permian marine sedimentary rocks in the foothills of the Takitimu Mountains, Southland. These include several leaf fossils of the classic Gondwanan plant *Glossopteris*, which is found abundantly in Antarctica, Africa, South America, Australia and India. This provides evidence for the existence, at that time, of the Southern Hemisphere supercontinent of Gondwana. Throughout the Paleozoic and most of the subsequent Mesozoic, present day South America, Africa, India, Australia and Antarctica were linked together as one giant continent (or plate). During all this time the New Zealand region was mostly under the sea, just off the east Australian coast of Gondwana. At times during the Cambrian and Permian, volcanic islands may have peppered the region and, in the latter half of the Paleozoic, parts of western New Zealand may have been uplifted to form the coastal margin of Gondwana.

During the Paleozoic the huge crustal plate that was Gondwana migrated slowly over the surface of the earth. In the Cambrian it straddled the equator with the New Zealand region in the Northern Hemisphere enjoying a warm climate. It progressively moved south so that in the Permian, Gondwana covered the South Pole region with New Zealand at about 50 degrees South and quite cool.

69. Some of the most common Permian fossils in New Zealand are bivalves, such as this 7cm specimen of *Trabeculatia trabeculum* (with both valves still joined) from Whangamoa River, Nelson.

70. A 30cm slab of Permian mudstone containing numerous leaf impressions of *Glossopteris*. This distinctive plant has been found as fossils in all the southern hemisphere lands that were formerly joined together as the supercontinent of Gondwana.

Trilobites – extinct sea-lice

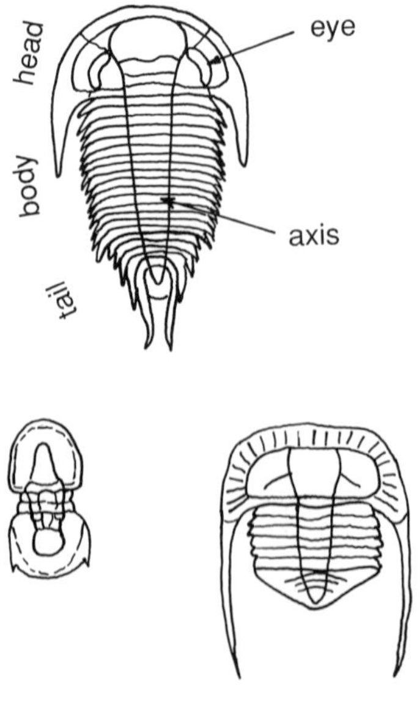

71. Some of the variety of shapes of fossil trilobites.

Finding trilobites is the dream of many fossil hunters. Trilobites have this fascination, possibly because they are an extinct group and were the first animals to develop hard parts that could be fossilised. For most people trilobites are the oldest fossils they will ever find. The oldest trilobites define the beginning of the Paleozoic era. They rapidly diversified into the many forms present in the Cambrian and Ordovician periods. Their variety began to decline in the Silurian and only a very few forms continued through to the Permian. All the remainder became extinct at the end of the Paleozoic.

New Zealand's oldest fossils are middle Cambrian (530 million years old) trilobites from Cobb Valley, northwest Nelson. A few poorly preserved trilobites have also been found elsewhere in northwest Nelson and are of Ordovician (Mt Patriarch) and Devonian (Baton River) age. A few Permian trilobites have been collected in western Southland.

Like their relatives, such as the insects, wood-lice and crayfish, trilobites belong to the Phylum Arthropoda and have a hard chitinous external skeleton. In trilobites the dorsal (upper) part of the skeleton has been hardened with calcite and is frequently fossilised. Trilobites are divided into head, body and tail regions and also longitudinally into three lobes (hence tri-lobite), with a central axis flanked by two lateral areas. The body and tail are segmented. Each segment had a pair of legs, although these are seldom preserved. Trilobites crawled and swam around on or near the floor of shallow seas and some may even have burrowed in the soft sea-floor sediment. Many trilobites have a pair of compound eyes on the head shield, although others are blind. A wide variety of body shapes are found in overseas trilobites but only a limited number of forms are so far known from New Zealand.

72- 74. Broken portions of three Cambrian trilobites from Cobb Valley. Left: a dolichometopid head shield. Centre: an agnostid head shield (blind). Right: a tail of *Dorypyge*. Photos: Roger Cooper. (each 0.5 - 1cm across)

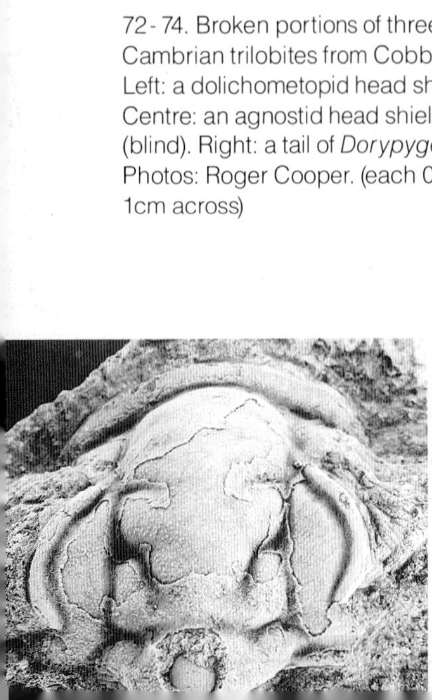

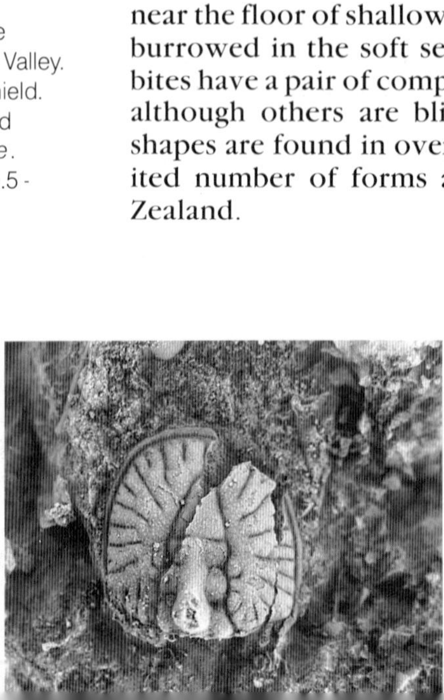

75. These graptolites are the carbonised compressions of ancient colonial animals that lived in the ocean plankton. They are most commonly found as faint white markings on the splitting planes of black slates, such as these Ordovician specimens from Cobb Valley, northwest Nelson. (slab 12cm)

76. Graptolites evolved into a wide variety of branched and unbranched forms. Colonies average 1 - 10cm in size. These Ordovician forms are all from Aorangi Mine, northwest Nelson.
Top: a tetragraptid.
Centre: *Pseudoisograptus*.
Bottom: *Goniograptus*.
Photos: Roger Cooper.

Graptolites – ancient plankton

Graptolites are an extinct group of animals most commonly seen fossilised in black shales or slate. They appear like faint white markings, as if drawn with a pencil, hence their name (graptos = inscribed). They first appeared in the late Cambrian, rapidly evolved into many forms in the Ordovician and Silurian, and became extinct by the Permian. In New Zealand they are only known from Ordovician rocks in northwest Nelson (eg. Cobb Valley), Buller and Fiordland (Cape Providence).

Graptolites consist of one or more branches, each with characteristic saw-tooth markings along the side. These saw-teeth are tiny cups, each containing a separate animal (or polyp), connected to others by a stalk. Each polyp had a pair of tentacles used for breathing and filtering out minute food particles from the sea water. Graptolites are branched colonial organisms that floated about in the plankton of the ocean and were widely distributed around the world by currents. Because of their world-wide distribution and their rapid evolution from one species to another, graptolites have proved very useful in identifying rocks of the same age in different continents.

Graptolites appear to be ancient hemichordates, primitive members of the Phylum Chordata, which includes vertebrates and humans.

Sea lilies – crinoids, carpoids and blastoids

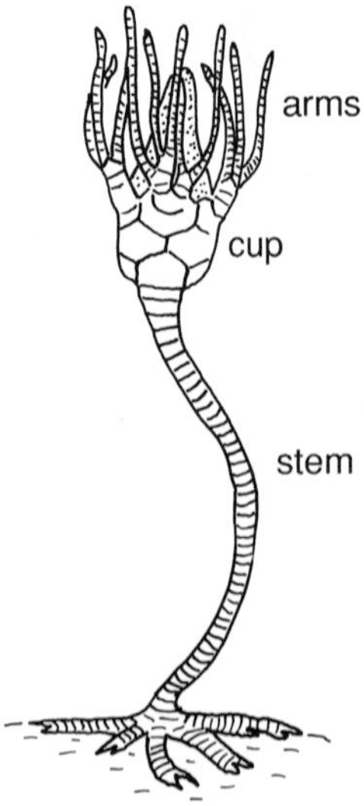

77. A typical sea lily (crinoid).

Fossil sea lilies (crinoids) and their relatives (carpoids and blastoids) are abundant in Paleozoic rocks overseas. In New Zealand only a handful of well-preserved specimens have been found, although the fragmented stems (ossicles) of sea lilies are commonly found in our Ordovician, Devonian, Permian and Mesozoic rocks. All three groups belong to the phylum Echinodermata of "spiny-skinned" marine animals. Their body wall consists of calcified plates that are frequently preserved as fossils.

Crinoids appeared in the Ordovician, had their peak abundance in the Paleozoic and have survived in limited numbers and variety through into today's seas. They consist of a globular or cup-shaped body with five basic feeding arms, sometimes branched, and a flexible stem attached to the sea floor. The cup has a network of external plates and the stem and arms are composed of stacks of circular or pentagonal disc-like plates, called ossicles.

Blastoids are similar to crinoids but did not survive beyond the Paleozoic era. Carpoids are also only known from the Paleozoic. They consist of a flattened, assymetric cup and a tail rather than a stem. They were free-living on the sea-floor rather than attached.

78. A rubber cast made from the flattened natural mould of a sea lily (crinoid) found in Devonian rocks near Reefton. The external plates of the cup are in the centre giving rise to numerous branched arms made of circular ossicles (below).

79. The only New Zealand fossil carpoid (kind of sea lily) is this 5cm specimen of *Allanicytidium flemingi* from Devonian rocks at Reefton.

Lampshells – brachiopods

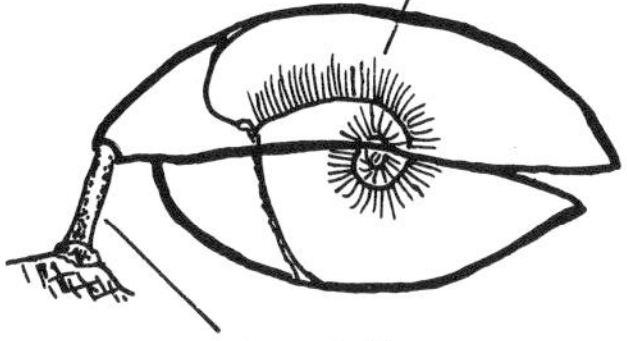

80. A generalised lampshell, showing the lophophore inside the shell.

Lampshells are the largest group of fossil animals. They constitute the phylum Brachiopoda and are commonly known as lampshells because of their resemblance to Roman oil lamps. Lampshells have a hard external skeleton consisting of two valves – upper and lower. The valves are usually of unequal size which helps distinguish them from bivalve molluscs.

Lampshells are sedentary, marine animals, 5mm - 8cm in size, usually attached by a muscular stalk to rock or large shells. A few cement one shell to rock and some ancient forms are thought to have rested freely on the soft sea floor. Inside the shells they possess a specialised organ called a lophophore for filtering out microscopic food particles from sea water. Small hairs on the lophophore beat to create currents in and out of the shell to assist filter feeding.

The oldest fossil lampshells appeared in the Cambrian and evolved into a wide variety of forms by the late Paleozoic. The number of forms decreased from Permian time onwards, although New Zealand is unusual in the richness of its fossil lampshells in the Cenozoic and the abundance of lampshells that still live around our coast, especially Stewart Island. New Zealand's Cambrian and Ordovician brachiopods are inconspicuous, but a rich diversity of forms are present in our Devonian and Permian rocks.

81. Fossil casts of the Devonian lampshell, *Maoristrophia neozelanica*, from Reefton, together with numerous circular sea lily (crinoid) stem ossicles. Each lampshell is about 4cm across.

82. A large (6cm) middle Miocene (14 million year old) lampshell, *Stethothyris epsilon*, from Clifden, Southland.

83. The Mesozoic Era is divided into three periods.

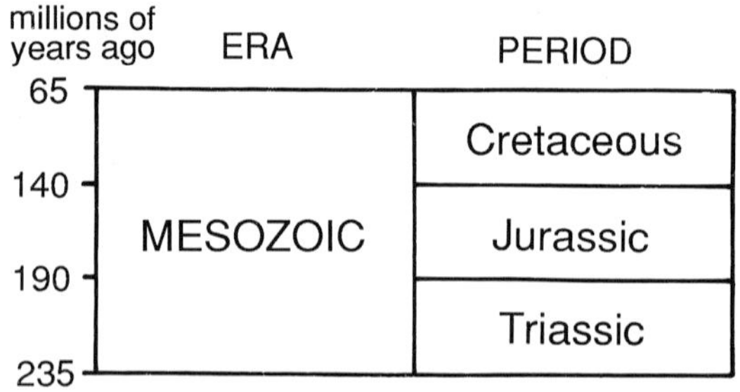

84. Location of Mesozoic rocks (shaded) in New Zealand.

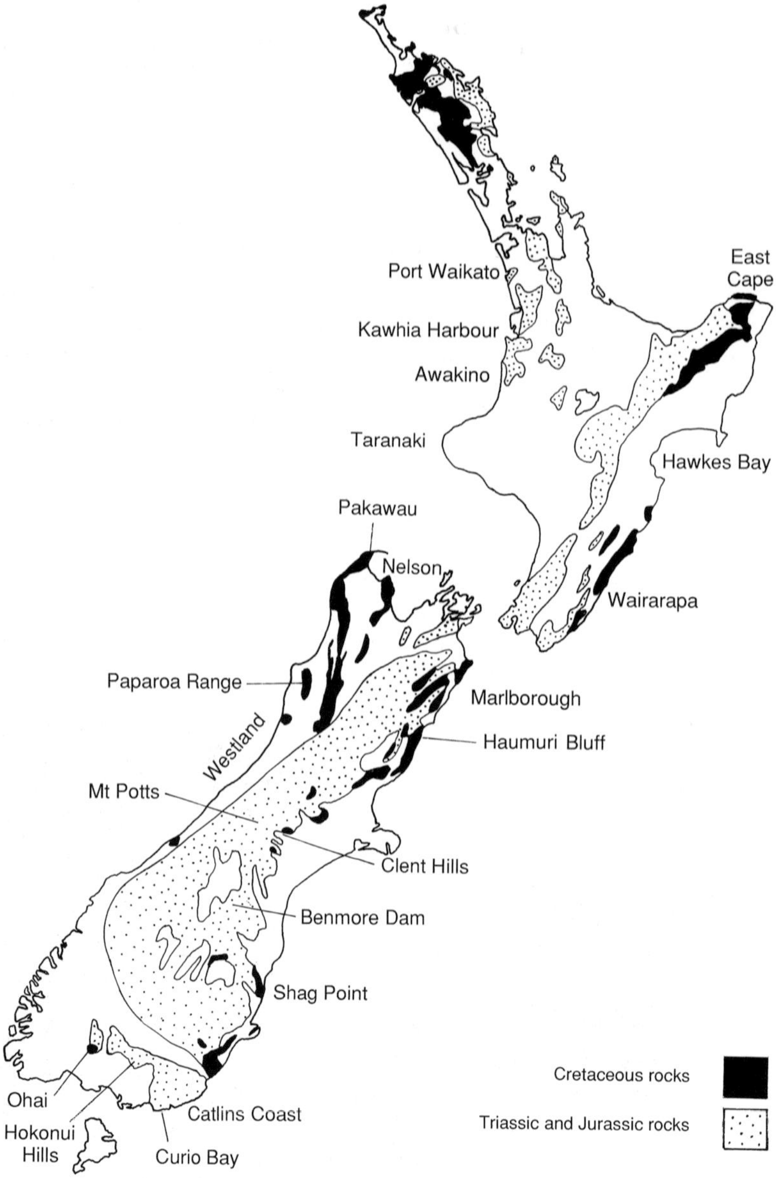

New Zealand's record of Mesozoic life – Age of Reptiles

The Mesozoic era consists of three periods and spans the time between two major episodes of mass extinction, 235 million years ago and 65 million years ago. It is commonly known as the age of reptiles, for during this time large and fierce-looking reptiles such as dinosaurs and sea dragons dominated life on land and in the sea.

Sedimentary rocks that accumulated during the Mesozoic are common in many parts of New Zealand, and include most of the hard greywacke rocks of the main ranges in both islands. Locally abundant fossils preserve an excellent record of the marine life which had hard parts and inhabited the seas of the New Zealand region at this time. Fossilised remains of plants and animals that lived on land are much less common, especially those of Triassic and Jurassic age.

The small number of sea animals that survived the mass extinctions at the end of the Paleozoic soon developed into a myriad of new forms in the Triassic period. Most abundant of those found fossil in our Mesozoic rocks are the bivalves or clams, followed by lampshells, belemnites, and less common ammonites, snails, sea eggs, sea lilies, bryozoa, worms and reptiles. Rocks rich in marine fossils of Triassic and Jurassic age are frequently found in a belt on the west side of the North Island between Port Waikato and Awakino, and also in northern Southland in a belt between Te Anau and the Catlins coast. Rocks rich in Triassic fossils also occur southeast of Nelson city.

85. Lampshells are frequently found fossilised in New Zealand's Mesozoic rocks, like this *Clavigera bisulcata* from the late Triassic of Marakopa coast, King Country.

During the Triassic the New Zealand region was mostly under the sea, off the south coast of the supercontinent of Gondwana. Areas that may have been land, like northwest Nelson, were the most southerly and coolest parts of the great continent and this may explain why no fossil remains of dinosaurs or their land-dwelling ancestors have been found in our Triassic and Jurassic rocks. Similarly ammonites lived abundantly in warm tropical seas, but their sparseness as fossils in New Zealand rocks is confirmation of our cool climate throughout most of the Mesozoic.

Throughout the late Paleozoic and Mesozoic the New Zealand region lay along the collision boundary between two great crustal plates of the earth – the Gondwanan and Pacific. Initially, during the Permian, Triassic and early Jurassic, a vast thickness of sediments (tens of kilometres thick) accumulated on the subsiding floor of the sea in our area and a chain of volcanoes or volcanic islands erupted intermittently nearby. These sediments were carried by earth movement (subduction) deep into the earth, where they were hardened into greywacke or metamorphosed into schist.

About 150 million years ago, towards the end of the Jurassic period, the downward movement of the Pacific plate became jammed beneath the leading edge of the Gondwanan plate. The collision between the two plates was translated into a buckling upwards along the boundary and much of the sea floor and underlying rocks were pushed up to form land in the New Zealand region. By the Cretaceous period a huge land, about half the size of modern Australia, had been formed. It extended from New Caledonia through New Zealand to the Chatham and Campbell Islands. Many Gondwanan animals and plants moved into the New Zealand region at this time. Examples include the ancestors of the tuatara, our endemic frogs, moas, kiwis, our large land snails and giant wetas.

Thus during the early Cretaceous almost all of New Zealand was land. Marine rocks and fossils of this age are virtually absent. By the middle Cretaceous, some 90 million years ago, Gondwana was beginning to break apart and rifts began opening up between Australia and New Zealand. During the latter part of the Cretaceous and into the early Tertiary (between 80 and 55 million years ago), New Zealand split away from Gondwana forming the Tasman Sea we know today and some of the southern ocean between us and Antarctica. Ancestral marsupials, land mammals and snakes evolved in Gondwana, in the late Cretaceous

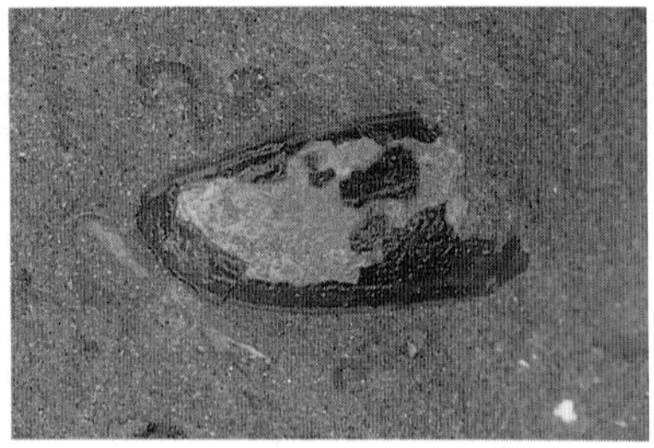

86. (Above) New Zealand's oldest insect fossil is this 1cm long wing case of a forest leaf-eating beetle found in late Cretaceous rocks, inland Hawkes Bay.

87. (Right) The 12cm long underbelly plate of a turtle from late Cretaceous rocks in inland Hawkes Bay. Occasional fossil remains of turtles are known from Cretaceous, Paleocene, Oligocene and early Miocene rocks in New Zealand. There are still odd strandings on our shores today.

88. Fossil snails are not abundant in New Zealand's Mesozoic rocks. This 5cm specimen of *Raha spectabilis* comes from late Triassic rocks of Roaring Bay, northeast Southland.

and early Tertiary, just too late to reach New Zealand before we moved away.

As New Zealand split away, parts of the region began to subside. In the area between East Cape and Wairarapa and south to Marlborough, sediment accumulated on the floor of the encroaching sea preserving abundant fossil bivalves, belemnites, lampshells and snails, together with occasional mosasaurs, plesiosaurs and rare dinosaurs, washed in from the adjacent land. The land was being worn down by erosion. Thick beds of coal, river silt and gravels accumulated, in large swamps and alluvial plains, in south Otago, Westland, northwest Nelson and offshore Taranaki. Many of these late Cretaceous rocks now exposed at the surface contain beds rich in fossil leaves, wood and occasional freshwater mussels.

Suddenly at the end of the Cretaceous, 65 million years ago, the largest known world-wide extinction event took place. It killed off all the dinosaurs, flying reptiles and sea dragons. All the ammonites and belemnites too became extinct. It greatly reduced the number and variety of almost all groups of animals and plants, whether they lived on land or in the sea. Indeed three-quarters of all living creatures are believed to have died out. There have been many postulated explanations over the years but the most likely cause was the collision of an asteroid (10-15km across) with the earth. This would have created thick dust clouds that encircled the globe, shutting out sunlight and thereby killing off many plants on land and in the sea. Animals that relied on these plants for food would have died from starvation, as would the predators that lived on them. Evidence of this major extinction at the end of the Mesozoic can be studied in the fossils preserved in rock sequences in Marlborough and the Wairarapa.

89. Shell beds containing abundant scallop-like bivalves of the genus *Monotis* are found in late Triassic rocks in New Zealand and many other parts of the world. They evolved rapidly and are particularly useful for dating and correlating rocks of the same age in different countries. This 25cm slab comes from Kiritehere on the King Country coast.

90. Trigonids are a distinctive group of bivalves. They are roughly triangular in shape, have two kinds of shell ornament and are frequently found in New Zealand Mesozoic rocks. Two shells of *Maoritrigonia nuggetensis* from Nugget Point, Southland, Triassic. (6cm across)

Fossil clams – bivalves

Bivalves are a large group within the phylum Mollusca. All possess two valves (shells) hinged together at the top and enclosing the soft parts. This group includes such well known delicacies as scallop, oyster, mussel, toheroa, cockle and clam. In addition to their two valves they also differ from other molluscs, such as snails, squid and chitons, by the lack of a head and the possession of a compressed, fleshy tongue-like foot that projects through the gape of the shells when it is feeding or moving.

All bivalves live in the sea except for freshwater mussels. Most are found on the sea floor as sluggish crawlers, burrowers in soft sediment, borers in rock or fixed by holdfasts or cement to solid objects. A few, such as scallops, are capable of swimming short distances.

The hard bivalve shells are readily preserved and fossilised. The oldest bivalves are found in middle Cambrian rocks (eg. Trilobite Rock) and they gradually evolved into a diversity of forms throughout the Paleozoic. By the Triassic, they had increased in abundance in the New Zealand region to succeed the lampshells as the dominant group of invertebrates living on the sea floor. Indeed throughout the Mesozoic, bivalves are the most abundant and important group of fossils in New Zealand. The evolving succession of bivalves are used as the basis for dividing up the Mesozoic era in New Zealand into 22 different time zones.

In New Zealand's Tertiary sedimentary rocks, bivalves remain abundant and jointly dominate the bigger fossils with their cousins, the sea snails (gastropods). Some fossil bivalves are also important in determining the age of Tertiary rocks in New Zealand, especially the rapidly evolving scallops and fanshells (pectinids) of the late Tertiary.

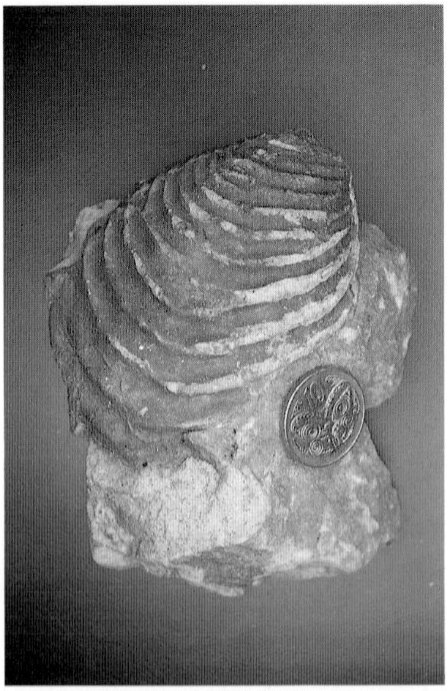

91. The bivalve genus *Inoceramus* is the most common and useful fossi in New Zealand's Jurassic and Cretaceous rocks. Its evolving succession of species forms the basis for recognising many of the 15 time divisions that these rocks have been divided into. *Inoceramus* in New Zealand was a strong-ribbed clam with some members growing as large as 1.5m across. This 6cm specimen is *Inoceramus australis* from the late Cretaceous of the Ureweras.

92. *Halobia lilliei* is a small, short-lived, scallop-like bivalve that occurs abundantly in late Triassic rocks in New Zealand. Its free-living larval stage allowed it to migrate right around the world at this time. (photo 7cm across)

93. The theropod dinosaur bone found in Hawkes Bay came from an animal looking something like the one on left.

94. The Hawkes Bay pterosaur bone came from a flying reptile that looked something like the creature below.

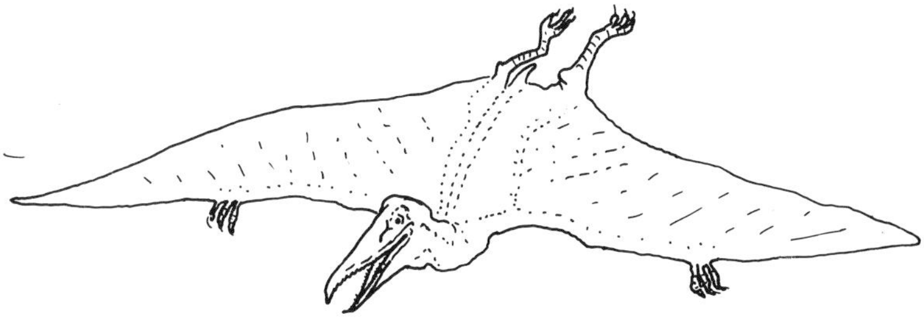

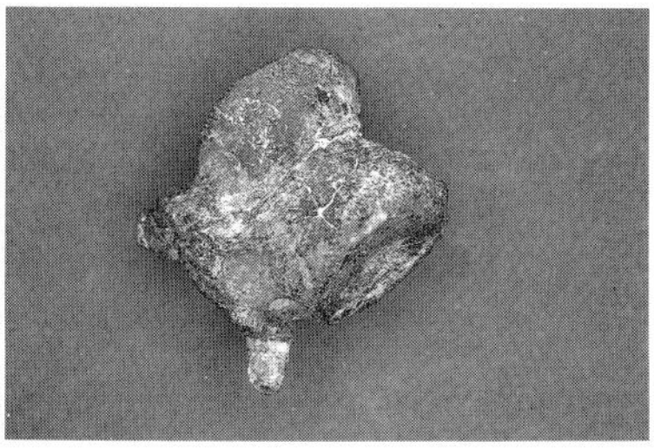

95. (Above) This 8cm bone is a tail vertebra from a two-legged carnivorous theropod dinosaur that stood about 2m high and was 7m long. This bone is one of only two individual dinosaur bones so far found in New Zealand. Both have been extracted from late Cretaceous (70 million year old) marine rocks in the Te Hoe River, Hawkes Bay. These dinosaurs lived on land and after they died their bodies were washed into the sea and their bones fossilised, together with many marine fossils.

Dinosaurs

Dinosaurs are children's favourite group of extinct animals and fossils. They evolved in the late Triassic, about 200 million years ago, and ruled the land during the Jurassic and Cretaceous, until their dramatic die out 65 million years ago at the end of the Mesozoic. Some dinosaurs were vegetarians and grew to huge sizes, whereas others were ferocious predators. Most lived in the Northern Hemisphere with few reaching southern lands.

Until the 1980s dinosaur and pterosaur (flying reptile) fossils were not known from New Zealand. In recent years a dedicated amateur paleontologist, Joan Wiffen, and her friends have made numerous exciting fossil discoveries in the late Cretaceous rocks of inland Hawkes Bay. These include mosasaurs, plesiosaurs, turtles, and single bones of a pterosaur and two dinosaurs – a carnivorous theropod and a 5m long herbivorous hypsilophodont.

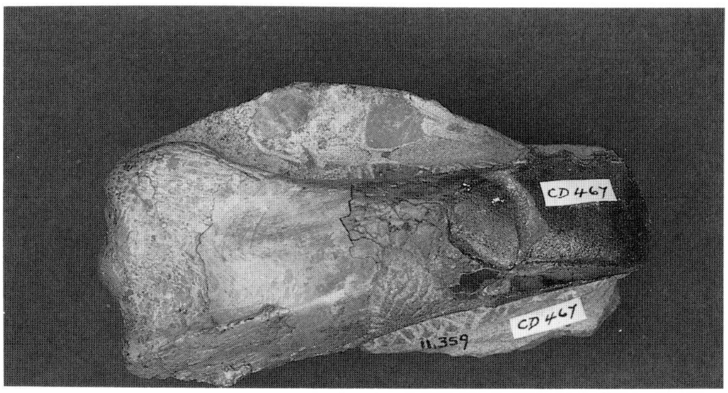

96. (Right) This light-weight bone is the only fossil remains so far found in New Zealand, of the giant flying reptiles, called pterosaurs. The bone comes from the same late Cretaceous (70 million year old) marine rocks in Hawkes Bay as our only dinosaur fossils. This 10cm long bone is from the wing of a medium-sized specimen with a 3m wing span. The largest pterosaurs overseas had 15m wing spans.

Sea dragons – ichthyosaurs, plesiosaurs and mosasaurs

Mesozoic seas were dominated by large swimming reptiles or sea dragons. The oldest group were the fish lizards or ichthyosaurs that flourished in the Triassic and Jurassic periods. They resembled large dolphins with numerous big teeth in their long jaws. Fossilised ichthyosaur bones have been found in Triassic rocks in inland Canterbury and the Hokonui Hills of northern Southland.

In the Cretaceous, ichthyosaurs were displaced in their dominance of the seas by mosasaurs (sea lizards) and plesiosaurs (sea serpents). Mosasaurs were up to 15m long and resembled modern monitor lizards. They were probably the most ferocious predators of Mesozoic seas. Plesiosaurs were of similar length to mosasaurs but had much broader bodies, long powerful paddle-like flippers and often a long neck. The legendary Loch Ness monster is ''modelled'' on a plesiosaur. Fossil bones and some spectacular complete skeletons of both plesiosaurs and mosasaurs have been excavated from late Cretaceous rocks in Hawkes Bay, Marlborough, North Canterbury and North Otago.

Like the dinosaurs, these giant sea dragons also died out at the end of the Mesozoic.

97. Fossilised bones of mosasaurs (top), ichthyosaurs (middle) and plesiosaurs (bottom) are found in New Zealand's Mesozoic rocks.

98. This long jaw with ferociously sharp teeth is part of a mosasaur skeleton collected from late Cretaceous rocks of Te Hoe River, inland Hawkes Bay.

99. The walls (septa) between the internal chambers in ammonoid shells are intricately folded and frilled as can be seen in this 12cm specimen of *Lytoceras taharoaense* from Kawhia (late Jurassic). The outer wall of one chamber has been painted white to show up the frills.

100. (Top right) The walls between the internal chambers in nautiloid shells are a simple curve as can be seen in this 10cm Miocene specimen of *Aturia coxi* from Little Totara River, southwest Nelson.

Ammonites and belemnites – shell-bearing squid

Ammonites and belemnites are two well known groups of extinct marine animals that are commonly found fossilised in New Zealand's Mesozoic rocks. Both belong to the group of Mollusca called Cephalopoda, which includes modern octopus, squid and cuttlefish. All cephalopods have a ring of tentacles or arms around the mouth and a funnel through which water can be squirted to help propel them through the water.

The oldest groups of fossil cephalopods (the nautiloids and ammonoids) had external, chambered shells with a large final chamber into which the animal could withdraw all its soft parts and tentacles. The nautiloids (which have lived from the Cambrian to the present day) differ from the ammonoids (Silurian to

Cretaceous) in the shape of the wall (septa) between chambers. In nautiloids the wall is a simple curve but in ammonoids it becomes folded and frilled towards the margin and this difference is readily seen in fossils. Nautiloids had their greatest variety of forms in the Ordovician and have been in decline ever since. They are now near extinction, being represented only by the modern pearly Nautilus. The first nautiloids had small, conical shells but as time passed these evolved into the flat, planispiral shells of all Mesozoic, Tertiary and modern forms. Occasional fossil nautiloids are found in New Zealand's Paleozoic and Tertiary rocks.

Ammonoids appeared in the Silurian and had four great bursts of evolution (Devonian, Carboniferous-Permian, Triassic, Jurassic-Cretaceous) when many forms developed. Each burst was followed by a major die out with the loss of a large part of the ammonoid fauna. Most ammonoids have tightly-coiled shells but short-lived straight or partly-coiled forms evolved on several occasions. During the Paleozoic, and into the Mesozoic, forms gradually evolved with more and more complexly-folded and frilled chamber walls. Those with the most intricately folded walls are the ammonites of Triassic to Cretaceous age.

Ammonites frequently form the nucleus inside concretions in the rocks. Scattered, usually poorly-preserved, Permian ammonoid fossils occur in Marlborough and Southland. Much better preserved Mesozoic ammonites have frequently been found around the Kaipara Harbour, Port Waikato, Kawhia Harbour, Wairarapa, Hokonui Hills and Catlins coast of northern Southland.

Belemnites are the ancestors of modern squid and cuttlefish and are distinguished from them by the presence of a hard, bullet-shaped, internal skeleton of calcite, called the guard, located inside the tail. Belemnites appeared in the late Paleozoic, but they are most abundant, overseas and in New Zealand rocks, in the Jurassic and Cretaceous. They became extinct by the early Tertiary. When the belemnite squid died, it sank to the sea floor, the flesh decayed and only the hard guard survived to be fossilised. They are common fossils in the Port Waikato, Kawhia, Wairarapa, Haumuri Bluff (Kaikoura coast) and Hokonui Hills areas.

Nautiloids, ammonoids and belemnites were mostly free-swimming animals that easily spread throughout the world's seas. They also evolved rapidly into many different, short-lived forms (especially the ammonites). Thus they are particularly useful in accurately recognising rocks of the same age world-wide.

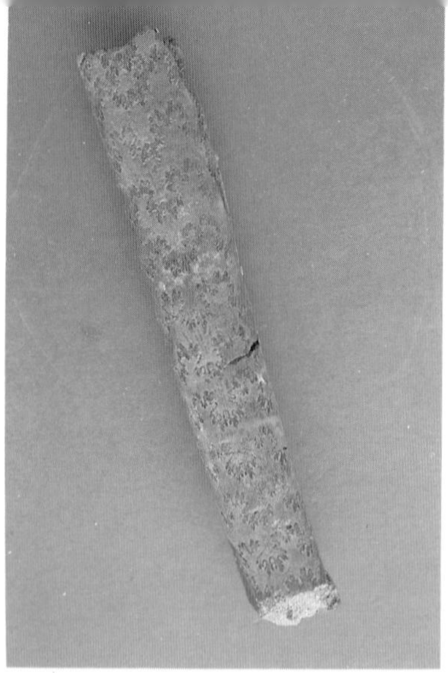

101. Most ammonites were tightly coiled in a single plane (= planispiral). During several of their bursts of rapid evolutionary radiation, high-spired, uncoiled and straight forms developed, such as this straight ammonite (*Baculites rectus*) from the Cretaceous of inland Hawkes Bay. (Length 8 cm)

102. These 6 -10cm long, bullet-shaped, belemnite guards were the internal skeletons of Mesozoic squid. Left: *Belemnopsis uhligi*. Middle: *Belemnopsis keari* (both late Jurassic). Right: *Dimitobelus superstes* (middle Cretaceous).

Mesozoic plants

The ferns which were so abundant on the New Zealand margin of Gondwana in the late Paleozoic were followed in the Mesozoic by common conifers, palms and later flowering plants. The main canopy trees in the Triassic and Jurassic forests were conifers and broadleaved ginkgos. Seed ferns, cycad palms, tree ferns, horsetails and lycopods comprised much of the understorey shrub layer. Judging by their abundant spores, mosses and liverworts were more common in the Mesozoic than they are today.

Major changes occurred in New Zealand's vegetation and worldwide in the Cretaceous with the appearance of the first flowering plants (Angiosperms). They evolved rapidly into many varieties and had become the dominant plant group by the late Cretaceous and have remained so ever since. By the end of the Cretaceous the ancestors of many modern plants had appeared. By this time the southern beeches (*Nothofagus*) and proteas had become common and diverse in New Zealand. Nevertheless two different groups of conifers, the podocarps and araucarians, remained common.

Throughout most of the Mesozoic, New Zealand's flora shared a lot in common with other Southern Hemisphere countries that were then all still part of Gondwana. Towards the end of the Cretaceous, New Zealand's plants began to evolve independently of other countries, as we split away from the supercontinent.

Our knowledge of New Zealand's Mesozoic plants comes mostly from study of their abundant fossil spores and pollen. Wood, leaves and seeds are less common but provide more graphic evidence of our ancient vegetation. Triassic plant macrofossils are mostly poorly preserved leaves and wood fragments. The best of these are found in rocks at Mt Potts and Benmore Dam (both South Canterbury).

Better preserved plant macrofossils come from New Zealand's Jurassic rocks, especially from Port Waikato (South Auckland), Curio Bay (Southland) and the Clent Hills (Canterbury). Common Cretaceous plant fossils can be readily found at places such as Pakawau (northwest Nelson), Paparoa Ranges (Westland), Malvern Hills (Canterbury), Shag Point (North Otago) and Ohai (Southland). Peat swamps were common features of the low-lying landscape of the late Cretaceous.

103. Most fossil ammonite shells are 5 - 20 cm in diameter, but a few giant forms have been found, like this Jurassic specimen from the Kawhia area, western North Island. It is New Zealand's largest ammonite at 1.42 m across and is on display in the New Zealand Geological Survey building in Lower Hutt.

104. The most common fossil fern in the New Zealand Jurassic is *Cladophlebis*, seen here beautifully preserved in hardened mudstone from Huriwai, Port Waikato. (Fern fossil is 6cm long)

105. (Above). The seed-fern *Dicroidium* occurs in Triassic rocks throughout ancient Gondwana. *Dicroidium odontopteroides* Tank Gully, Rangitata Valley, inland Canterbury. (3cm)

106. (Top right). Flowering plants appeared in the early Cretaceous and become a major group in New Zealand by the late Cretaceous. Broadleaved forms were most common. Leaves from late Cretaceous at Pakawau, northwest Nelson. (Leaf 12cm long)

107. (Mid right). Jurassic sandstone from Waikawa, Southland, rich with impressions of the ancient cycad *Ptilophyllum acutilobum*.

108. (Bottom right). Ginkgos were a major component of New Zealand's Triassic and Jurassic forests. *Ginkgophytopsis lacerata* Triassic rocks, Tank Gully, Rangitata Valley. (4cm)

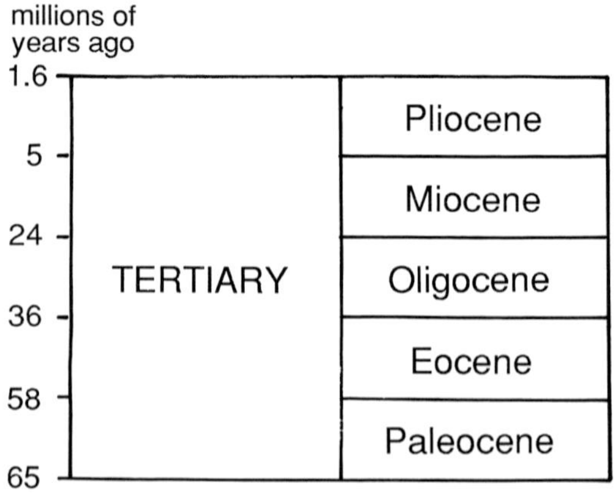

109. The Tertiary is divided into five periods.

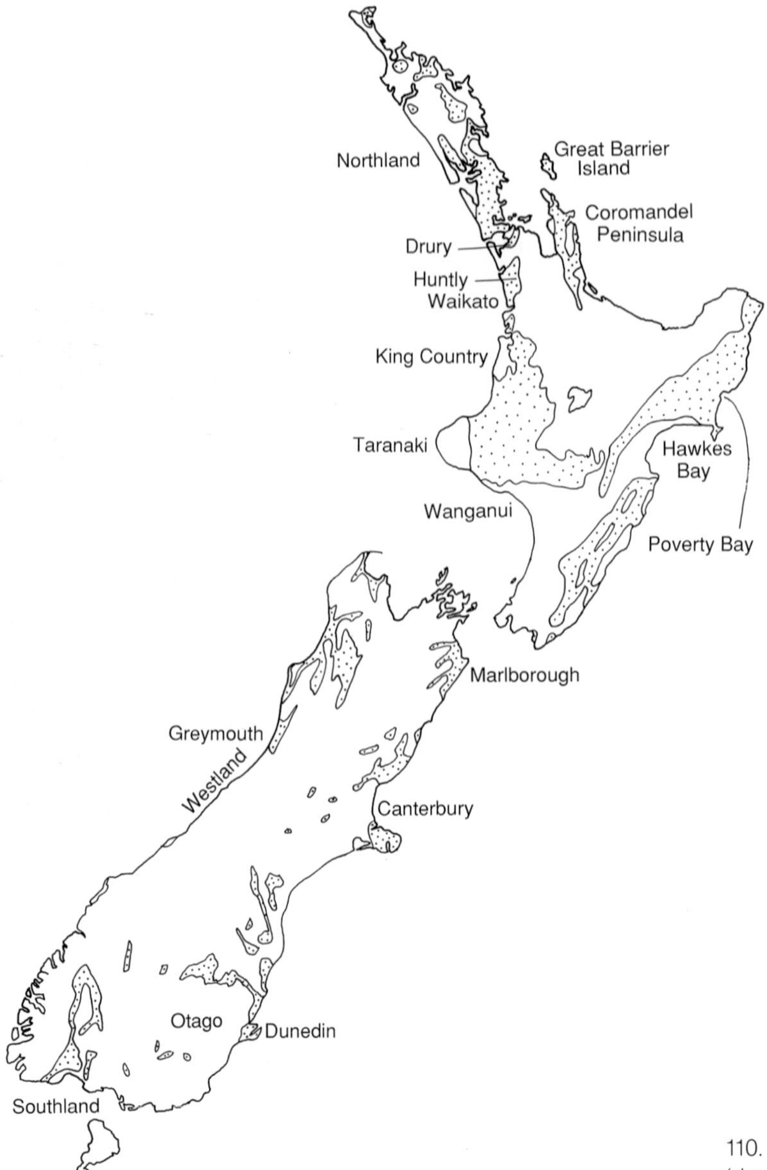

110. Location of Tertiary rocks (shaded) in New Zealand.

New Zealand's record of Tertiary life — Age of Mammals

The Tertiary consists of five periods and spans the time between the mass extinctions at the end of the Cretaceous, 65 million years ago, and the onset of a major period of ice ages in the Northern Hemisphere, about 2 million years ago. Overseas this interval is commonly known as the Age of Mammals, for after the extinction of the dinosaurs, mammals rapidly evolved to become the dominant creatures on land. New Zealand, having separated off from Australia and other continents in the Cretaceous, missed out on having its own land mammals.

We have very few fossil remains of terrestrial animals that provide direct evidence of life in New Zealand throughout the Tertiary. The composition of our modern fauna, and its relationships to overseas faunas and fossils, provides some assistance in piecing together a picture of Tertiary life. Birds dominated the land. Early ratites (moas) are believed to have existed prior to the breakup of Gondwana and were probably carried along as New Zealand split away. They diversified and would have been a major part of the Tertiary fauna (although our oldest known fossil moa bone is only 2 million years old). Other primitive endemic land creatures that lack a Tertiary fossil record, but probably accompanied the ratites as New Zealand split away include the tuatara, our native frogs, peripatus, giant wetas and possibly the large land snail *Paryphanta*.

The ancestors of most of our other insects, spiders, small land snails, lizards and birds, appear to have evolved in other countries and have arrived in New Zealand by chance at various times since the Cretaceous. Strong westerly winds frequently blow insects, spiders and land birds from Australia to New Zealand. Larger migratory birds can introduce eggs, larva and seeds in mud caked to their feet or feathers. Large pieces of driftwood occasionally are carried by storms and currents to New Zealand bringing immigrant lizards, snails and other small animals.

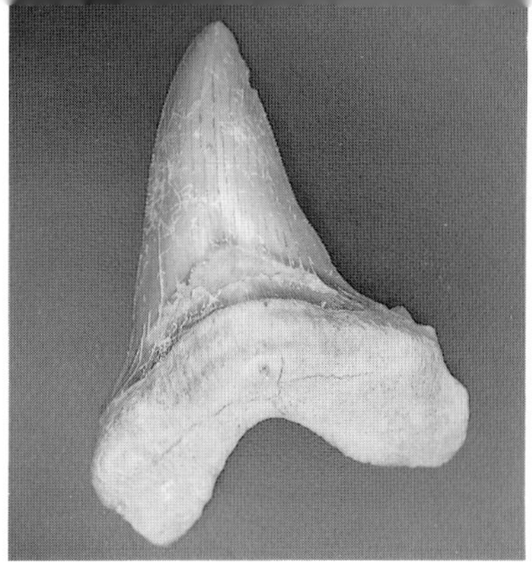

SHARKS

111. This 8cm fossil tooth from Oligocene (26 million year old) rocks in south Canterbury belonged to a giant 30m-long extinct shark (*Carcharadon*). Sharks, skates and rays have cartilagenous skeletons that are not normally preserved, but their sharp teeth are frequently found in New Zealand's Cretaceous and Tertiary sedimentary rocks.

112. These beautifully ornamented stalked barnacles (*Waiparaconus zelandicus*) come from Paleocene (60 million year old) greensand at Waipara, North Canterbury. Their unusually large and ornamented stalks are thought to have helped them anchor themselves in the soft mud of the seafloor. (Photo 12cm across)

113. The fossilised skeletons of extinct fish, such as this 4 million year old specimen from inland Hawkes Bay, are extremely rare finds in New Zealand rocks.

114. These three fossil barnacles (*Balanus decorus*) are still attached to an oyster shell. They were found in soft Pleistocene (1.5 million year old) rocks in Hawkes Bay.

BRYOZOA

115. A 5cm fossil bryozoan (*Retepora* — lace coral) from early Miocene (20 million year old) rocks at Cape Rodney, Northland. The phylum Bryozoa (moss animals) is a large but poorly known group of mostly marine colonial animals. They consist of numerous minute polyps (less than 0.5mm in size), each with a horny or calcareous cup, that join together to form encrusting or erect, cup-shaped, net-like or branched colonies, a few millimetres to 10cm in size. They are very common in New Zealand Tertiary rocks and are often a major constituent of our limestones.

116. Fossil freshwater mussels are frequently found in mudstone associated with Tertiary coal deposits. These impressions of *Hyridella aucklandica* occur in late Miocene lake sediments on Great Barrier Island.

117. Bivalves and snails are the most abundant macrofossils found in New Zealand's Tertiary rocks. This middle Miocene (15 million year old) marine bivalve, *Bassina speighti*, comes from Oparara, north Westland. (Width 4cm)

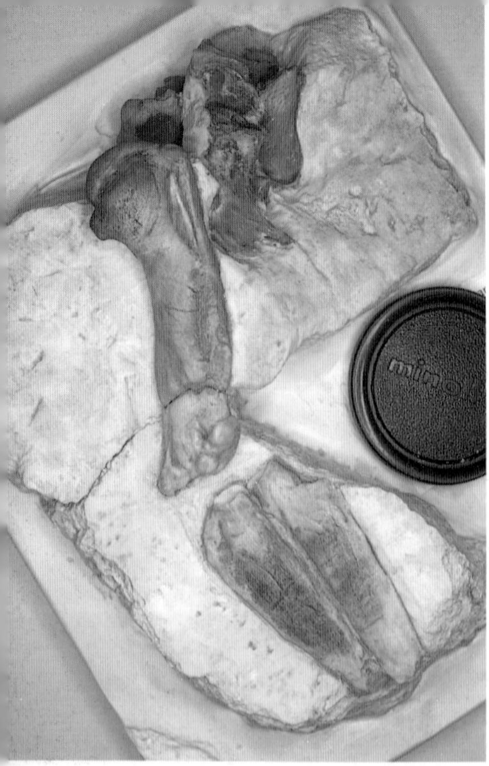

PENGUINS

118. The fossilised wing bones of an extinct Oligocene (30 million year old) penguin (*Platydyptes novaezealandiae*) from the Waitaki Valley. Penguins are a southern hemisphere group of birds and their New Zealand fossils provide the best record of their evolution. A variety of bones and partial skeletons have been found in Oligocene and late Eocene rocks of the South Island, with very few from the North Island. The oldest known fossil penguin is a primitive form found in early Eocene (50 million year old) rocks from Canterbury. By the late Eocene and Oligocene the penguins had diversified with fossils of at least 13 species from New Zealand. Most of these are large species and include the world's largest penguin (*Pachydyptes ponderosus*) which was as tall as an average person (1.6m) and weighed about 100kg. Since the Oligocene the penguins have slowly evolved into their modern forms.

Early in the Tertiary, the ancestors of our short-tailed bats, our kiwis, and three endemic families of primitive birds – the wattle birds, wrens and native thrushes – are thought to have arrived in New Zealand from Australia. Ancestors of our other birds and the long-tailed bat are believed to have arrived more recently, within the last 20 million years. It seems reasonable to assume that there were many extinct land creatures that lived in ancient New Zealand in the early Tertiary, but we know nothing about them because of the lack of fossil remains.

The record of plant life is well recorded, however, in rich sequences of fossilised pollen and spores and in a number of swamp and lake deposits, which contain well-preserved leaves and silicified wood. New Zealand's richest and most abundant fossil horizons however are of Tertiary shellfish and other marine animals that were buried in sediments that accumulated on the floor of the sea. Many of these are in moderately soft mudstone or sandstone and contain beautiful, well-preserved specimens that can be collected easily with a cold chisel. Fossils in Tertiary limestone are often cemented into the rock and more difficult to extract. The abundance of Tertiary marine fossils is an indication of the vast areas of modern New Zealand that were under the sea during some or all of this time.

The gradual encroachment of the sea over the land that began in the New Zealand region in the late Cretaceous continued through the Paleocene, Eocene and into the Oligocene. Throughout this time New Zealand was not part of a collision zone between the great crustal plates and, consequently, little if any land was being pushed upwards. Erosion reduced the mountains to low rolling hills and extensive river and coastal plains. Gentle subsidence, often in fault-bounded troughs, helped drown the land.

Some of these troughs filled with river and swamp sediment and peat that has since hardened to provide the coal measures of Northland, Waikato and Greymouth. Leaf fossils are frequently found in these rocks. Similar sediment, rich in vegetable matter, accumulated in troughs along the west coast of what is now the North Island and are the source of Taranaki's oil and gas reserves. Marine Paleocene and Eocene sediments mostly accumulated in eastern New Zealand, from East Cape to Otago. These contain sporadic and sometimes locally-abundant fossils, mostly snails and bivalves.

By the Oligocene, the sea had inundated over three quarters of the area of modern New Zealand. What little land was left was low-lying and supplied very little sediment to the surrounding seas. Since very little sand or mud was accumulating on the sea floor, vast areas of shell debris built up and blanketed the seabed throughout the New Zealand region. These shell deposits have since hardened and are the origin of the majority of New Zealand limestones. In areas of shallow Oligocene seas, in east Northland, south Auckland, King Country, Westland, Canterbury and Otago, the limestones are composed of broken and whole shells of bivalves, sea eggs, bryozoa (moss animals), brachiopods, barnacles, large foraminifera and calcareous algae. These extensive shallow seas were a major centre for the evolution of whales, dolphins and penguins in the Oligocene. In areas that had subsided to greater depths, such as west Northland, Poverty Bay, Hawkes Bay, Marlborough, North Canterbury and offshore Taranaki, the limestones are fine-grained and composed largely of plankton microfossils (calcareous ooze), such as foraminifera and coccoliths.

During the Eocene, Oligocene and early Miocene many marine creatures reached New Zealand in ocean currents, as eggs, larvae or adults. Some became established and gave rise to new endemic forms. The early and middle Miocene was the warmest period during the Tertiary and fossils of this age show that many tropical animals and plants, such as reef corals, cone shells, palm trees and coconuts, colonised and flourished in northern New Zealand.

During the late Miocene and Pliocene, the Antarctic ice cap was formed and world temperatures began to drop. Cooler climate animals and plants moved northwards through New Zealand at this time and many warmer climate forms were pushed northwards by the cold and disappeared.

In the late Oligocene and early Miocene the collision zone between the Pacific and Indian Plates, which passes through New Zealand today, was initiated. The resultant earth forces cut the land into fault blocks; some were pushed upwards to form new land and others sank to form deep, sea basins. Large volcanoes began erupting along the plate boundary in northern New Zealand.

Much of Southland and Otago was little affected, however. There a large area remained as low-lying plain and swamp, where sediment and plant material accumulated. This became the source of vast lignite coalfields found in these places today.

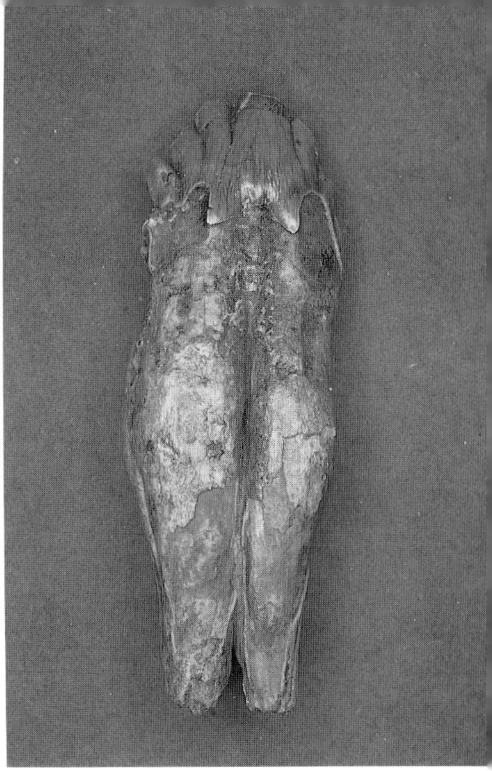

WHALES

119. The 8cm long fossil tooth of a primitive toothed whale (*Kekenodon onamata*), from Oligocene (30 million year old) rocks, Waitaki Valley. Cetacea (whales, dolphins and porpoises) evolved from land mammals in the early Eocene and filled the niche left vacant by the extinction of the sea dragons at the end of the Mesozoic. Over the years a wide variety of Cetacean skeletons and bones have been found in Oligocene rocks of Otago and Canterbury. These document the evolution and appearance of our modern groups of toothed whales, dolphins, porpoises and baleen whales.

120. Starfish usually decay and break up quite quickly when they die and as a result are rarely found fossilised. One of the few starfish fossils from New Zealand is this 10 cm asteroid (*Proserpinaster canopus*) from Miocene sandstone in the Mata River, East Cape.

121. Fossil pieces of the branching colonial coral, *Platyhelia distans* from middle Miocene (12 million year old) rocks at Tokomaru, East Cape. Here the cups of individual polyps have fused together to form a colony. (Length 5cm)

122. This near perfect specimen of a 30cm long fossilised crayfish (*Jasus flemingi*) comes from early Miocene (18 million year old) rocks in the Tarakohe cement quarry, northwest Nelson. Crayfish fossils are only occasionally found in New Zealand's Jurassic, Cretaceous and Tertiary rocks. Fossils of small freshwater crayfish have also rarely been collected.

By the middle Miocene most of northern New Zealand had been pushed up out of the sea and mountains began to form in central New Zealand. Both east and west coasts of the North Island remained under the sea accumulating considerable thicknesses of fossil-bearing sedimentary rocks.

In the Pliocene and Quaternary earth movements have been even more severe. These have mostly affected a belt along the plate boundary, running through the east and central North Island and the north and west South Island. During this period, the mountain ranges have been pushed up throughout the belt and most of the lateral movement has occurred on the Alpine Fault. Large thicknesses of marine sediment accumulated, in rapidly sinking areas in Taranaki, Wanganui and the North Island east coast, before themselves being pushed up to form land. Thus in these areas Pliocene and Pleistocene shellfish fossils are frequently found in soft "papa" rocks and can often be collected using a pocket knife. Relatively soft, bluff-forming limestone throughout Hawkes Bay contains many whole fossil barnacles and shells. These accumulated as shallow sea floor shellbanks during the Pliocene period.

123. Fossils provide information about past environments which is essential in reconstructing a picture of the past geography of a region. These maps show how the shape and form of the New Zealand area has changed during the Tertiary.

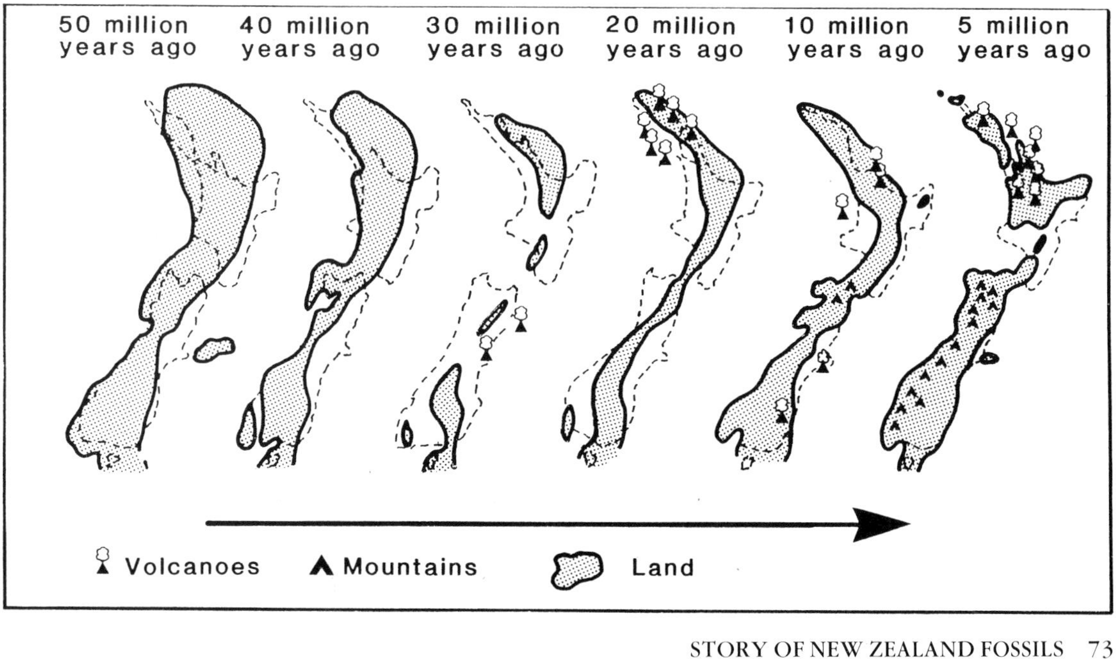

Snails and tusk shells – gastropods and scaphopods

Snails and tusk shells are two groups of molluscs, and are related to the bivalves (such as clams and oysters), cephalopods (ammonites, belemnites, squid) and chitons.

The snails, or gastropods, mostly have a single, usually spiral-coiled shell, a large broad foot and a well-defined head with eyes and tentacles. They first appeared in the Cambrian and slowly increased, in abundance and variety, through the Paleozoic and Mesozoic, reaching their peak of development in the Tertiary and Quaternary. In New Zealand, fossil gastropod shells are uncommon in Paleozoic and Mesozoic rocks but, together with bivalves, are the most abundant fossils found in our Tertiary marine rocks. Here some of the more rapidly evolving groups, such as *Struthiolaria*, are particularly useful in determining the age of rocks.

Most gastropods lived on the floor of the sea and had a spirally-coiled shell in a variety of shapes and sizes. A few evolved lungs and successfully colonised the land, like our large flax and kauri snails. Others became adapted to living in fresh water and another group, the pteropods, developed a winged foot for swiming and became part of the plankton of the oceans. The conservative group, known as limpets, never developed a coiled shell. The sea slugs lost the shell or have retained a reduced internal shell. Very few fossil freshwater snails or land snails are known from New Zealand, although very young landsnail shells (up to a few thousand years old) are frequently found in caves and prehistoric middens.

Tusk shells, or scaphopods, can be recognised by their slightly curved, tubular shells that are open at both ends. The animal lived partly buried in sand on the floor of the sea with its foot extended out through the bottom and siphons through the narrower top. Tusk shells are known as fossils from the Silurian through to the present day, but in New Zealand fossil examples first appear in our Triassic rocks. They are usually uncommon but can be locally abundant in Miocene rocks.

124. A middle Miocene (15 million year old) marine snail, *Coluzea kiosk*, from Clifden, Southland. (Length 12cm)

125. *Conus hendersoni* is a Miocene fossil belonging to the tropical group of gastropods known as cone shells. During the middle Tertiary (10-30 million years ago) many tropical gastropods occur in our fossil record. Their larvae probably arrived here in ocean currents and settled in the warm seas that bathed our coasts at that time. (Specimen 4cm high)

126. Fossil limpets, *Emarginula haweraensis*, from soft Pliocene (3 million year old) rocks at Waihi Beach, west of Wanganui. (Each 2cm long)

127. Tusk shells (scaphopods), such as this Miocene (10 million year old) *Dentalium otamaringaense* from north Taranaki, are ocassionally found fossil in our Mesozoic and Tertiary sedimentary rocks. Their curved, tubular shells may be smooth or longitudinally ribbed and range in size from 1-12cm.

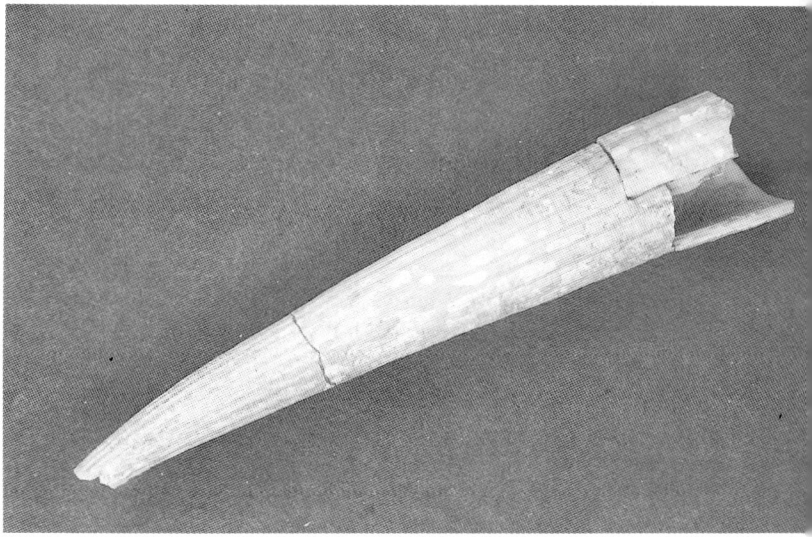

Corals

Corals belong to the phylum Coelenterata which also includes jellyfish, hydroids, anemones, sea lilies and gorgonians. Corals have an external, calcareous, cup-shaped skeleton in which an anemone-like polyp lives. Its feathery tentacles comb the water for plankton to eat. In colonial corals, adjoining cups join together and some may grow into major coral reefs. The calcareous skeletons are readily preserved and fossilised in marine sedimentary rocks.

Two extinct groups (rugose and tabulate corals) first appeared in the Ordovician and are extremely abundant and diverse in Paleozoic rocks overseas. Only a few forms are known in New Zealand's Ordovician, Devonian and Permian rocks. Modern corals (Scleractinians) evolved in the Triassic and occur in New Zealand Mesozoic and Tertiary rocks, although they are not common until the middle Eocene.

Most Tertiary coral fossils in New Zealand are solitary cups or branched colonial forms that did not build reefs. Solid reef-building corals are known from Permian and early Miocene rocks in northern New Zealand and indicate near tropical, shallow water conditions, then.

Gorgonians are a group of coral-like coelenterates that have long slender-branching colonies. Individual pieces of their calcareous or horny internal skeletons are frequently found fossilised in New Zealand's Tertiary sedimentary rocks.

128. Gorgonian corals, like *Melitodes hamiltoni* (early Miocene, Kakanui), have internal skeletons composed of small solid rods. (Total length 3cm)

129. Specimens of the solitary cup-shaped coral *Stephanocyathus mantelli* (3cm diameter) from early Miocene (20 million year old) rocks, near Timaru. Within the cup, corals have a distinctive pattern of radiating vertical walls called septa.

Sea eggs and starfish – echinoderms

A wide variety of echinoderm forms lived in the Paleozoic but only four groups – sea eggs, starfish, sea cucumbers and sea lilies survived into the Mesozoic and still live today. Of these, the sea eggs (or echinoids) are the most commonly found fossilised. The shell is composed of calcite plates covered in spines. When a sea egg died, the spines fell away often to be fossilised separately. The soft parts decayed and the empty round, disc-like or heart-shaped shell filled with sediment and was often fossilised whole, preserving the beautiful pattern of pores, plates and tubercles.

130. The Spatangoids are a group of advanced heart-shaped sea eggs that evolved from the radially symmetrical ones in the Cretaceous. They are deposit feeders that burrowed in the soft sediment of the sea floor. Occasionally beautifully preserved specimens are found fossilised in New Zealand's Tertiary rocks, like this 8cm, *Pericosmus crawfordi* (Miocene, Clifden, Southland).

Sea eggs first appeared in the Ordovician. Modern-looking forms developed in the Triassic and Jurassic with the most advanced group, the heart-shaped Spatangoids, evolving in the Cretaceous. Broken or distorted sea egg fossils are quite common in New Zealand Tertiary rocks and occasionally a perfect specimen is found.

The starfish and brittlestars (Asterozoa) are known from rather sporadic fossils and first appeared in the late Cambrian. Only a handful of good starfish fossils (all of Cretaceous or Tertiary age) have been found in New Zealand.

Sea cucumbers or Holothurians have an internal skeleton of separate calcite spicules. Scattered microscopic spicules are our only fossilised record of this group, which first appeared in the Ordovician.

131. A 5cm fossil sea egg, *Histocidaris mackayi*, from Oligocene rocks, north Otago.

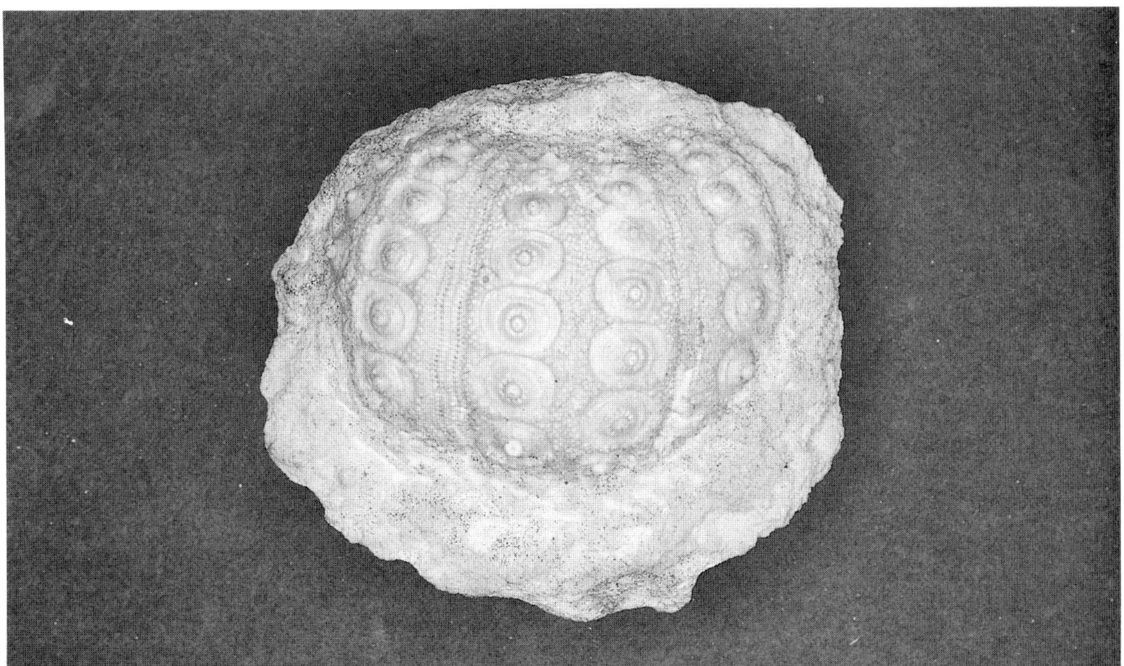

132. A middle Miocene (12 million year old) fossil crab (*Tumidocarcinus giganteus*) from Dovedale Stream, North Canterbury. (12cm across)

Crabs, crayfish and barnacles

Crabs, crayfish and barnacles all belong in the Phylum Arthropoda, together with other invertebrates that have chitinous external skeletons and jointed limbs, such as insects, spiders, slaters, trilobites and sand hoppers.

The solidly built external shells and claws of crayfish and crabs are sometimes preserved as fossils. In a few places, such as Motunau and Glenafric in North Canterbury, complete crab fossils are commonly found inside baseball-sized concretions. Both groups first appeared in the Jurassic and slowly evolved into a variety of forms, a number of which have been found in New Zealand's Jurassic, Cretaceous and especially Tertiary rocks.

Barnacles are marine animals that live inside a protective shell composed of a number of calcareous plates. Stalked barnacles attach themselves to rocks, shells or driftwood with a flexible stalk, whereas fixed barnacles cement their more solid shell directly to the substrate. When a barnacle dies, the soft parts decay and the protective plates may stay intact as an empty shell, or break apart before being preserved as fossils. The earliest barnacles are Silurian. They evolved slowly until the early Tertiary when many new forms appeared, including the ancestors of our modern fauna. Fossil barnacles are rare in New Zealand Mesozoic rocks but quite common, especially as individual plates, in our Tertiary rocks.

133. This 6cm long plate is one of six that comprised the shell of a large, deep water barnacle (*Bathylasma aucklandica*). It comes from early Miocene (20 million year old) rocks on Motutapu Island, Auckland.

134. The southern beeches of modern New Zealand all have relatively small leaves, but the dominant forest trees in the Eocene, Oligocene and Miocene were a group of large-leaved beeches (the *brassi* group), such as Nothofagus oliveri from late Miocene (10 million year old) rocks near Murchison. Similar large-leaved beeches still grow in warm, high humidity conditions in the mountains of New Guinea and New Caledonia, but they died out in New Zealand during the cold climate of the Ice Ages. (Leaf 5cm long)

Tertiary plants

Our record of New Zealand's terrestrial vegetation in the Tertiary is mostly obtained from the rich assemblages of fossil pollens and spores that occur in our sedimentary rocks (both non-marine and marine). While fossil leaves and wood are moderately common, they have been little studied and provide only minor additional information.

Podocarps and proteas dominated our Paleocene flora but declined in the Eocene. The relatively warm Eocene climate allowed palms such as the mangrove palm (*Nipa*) to flourish in parts of the country. Southern beeches (especially the warmth-loving large-leaved *brassi* group, but also some of the smaller leaved *fusca* types) were the dominant forest trees throughout New Zealand from the Eocene until the end of the Miocene. The somewhat cooler forests of low-lying New Zealand in the Oligocene and early Miocene included (in addition to beeches) prominent Casuarinas, Myrtles (eg. *Leptospermum*, *Metrosideros*) and podocarps.

In the early and middle Miocene, Northland was almost tropical with an abundance of palms, including a small-fruited coconut. Miocene vegetation became more diverse as earth movements produced greater topography and a wider variety of forest types resulted. As the mountains rose in the middle Miocene to Pleistocene, our distinctive alpine flora developed. Some alpine plants evolved from lowland ancestors

135. Fossil seed of a tropical palm found in 15 million year old rocks, near Murchison. (Seed diameter 2cm)

already in New Zealand and others arrived by dispersal from mountainous areas in Asia and South America.

Declining temperatures in the late Miocene, Pliocene and Pleistocene resulted in the extinction of many warmth-loving plants, such as *Casuarina*, *Acacia*, numerous palms and the large-leaved beeches. The major components of our modern flora took over.

Origins

A number of groups, like the southern beeches and some conifers, evolved in New Zealand from Cretaceous ancestors that they shared with other parts of Gondwana. A large portion of our terrestrial flora however, has evolved from ancestors that arrived in New Zealand during the Tertiary as seed carried by wind, sea or migratory birds from Australia, South-East Asia and other far flung lands. Our fossils record these arrivals. Plants such as *Leptospermum* (teatree) and *Metrosideros* (rata, pohutukawa) appeared in the Paleocene; *Phyllocladus* (tanekaha), *Phormium* (flax) and *Dracophyllum* (grass-trees) appeared in the Eocene; *Fuchsia* and *Coprosma* appeared in the Oligocene; and Compositae (daisies), *Pittosporum*, *Cordyline* (cabbage tree), *Macropiper* (kawakawa) and *Dysoxylum* (kohekohe) appeared in the early Miocene.

136. The fossilised frond of a palm similar to nikau, found in late Miocene (12 million year old) rocks near Murchison. The abundance of fossil palm pollen indicates that tropical and subtropical palms were a prominent part of New Zealand's vegetation throughout most of the Tertiary. Many died out as temperatures dropped with the onset of the Ice Ages about 2 million years ago. Photo: Dallas Mildenhall.

Macrofossils

Well preserved Tertiary leaf fossils are commonly found in mudstone within coal measure sequences, such as at Drury, Huntly, Greymouth, Central Otago and northern Southland. Other rich beds occur in freshwater mudstone, within volcanic rock sequences, on Great Barrier Island, Coromandel Peninsula and near Dunedin.

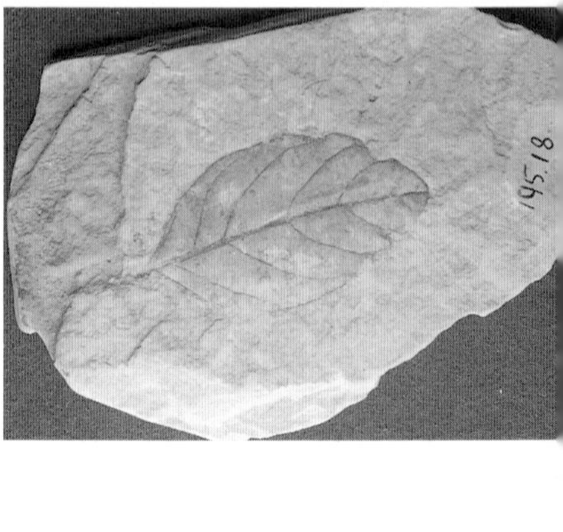

137. The well preserved impression of a Pliocene (3 million year old) fossil leaf of *Parsonsia obtusa* on a bedding plane in mudstone from Ormond, Poverty Bay. (Length 2.5cm)

138. An exquisite, 7cm fossil marine snail, *Murexul octogonus*, from soft Pleistocene rocks (700,000 years old) at Castlecliff, Wanganui.

New Zealand life in the Quaternary – Ice Ages

The Quaternary era spans the last 1.6 million years right up to the present day and has traditionally been taken to cover the period of the Great Ice Ages. Detailed research in the last 20 years however, has shown that major world cooling began in the Miocene with extensive Antarctic ice sheets developing during the period 12 to 5 million years ago. Signs of marked cooling in New Zealand and the Northern Hemisphere appear in the late Pliocene about 2.2 million years back. Since that time there have been at least 20 world-wide glacial (ice age) "episodes" separated by warmer interglacial episodes, or one cycle of cold and warm every 100,000 years approximately. Thus the Quaternary contains most of Ice Age time. It is divided into the Holocene period (last 10,000 years since the end of the last ice age) and Pleistocene period (all the rest).

The succession of cold and warm periods had a profound effect on New Zealand geography and nature. Between ice age periods, New Zealand's wildlife and plants were very similar to what there was 1000 years ago when humans first arrived. Glaciers and permanent snowfields were confined to the highest mountains and forests covered the low-lands and hills all the way to Stewart Island.

During each glacial episode, much of the world's water became frozen on land, especially in the Northern Hemisphere ice sheets. As a result, sea level fell 100-180m below present and vast areas of sea floor around New Zealand were exposed as land. The North Island, South Island and Stewart Island were linked together by dry land. The severe cold forced plants and animals to migrate northwards to warmer environments. Many warmth-loving kinds died out. Forest was confined to areas north of Taupo and a few coastal patches further south. The mountains were covered in permanent snow and glaciers. The foothills and low-lands were clothed in alpine grasses, rushes and sedges.

Fossils of Quaternary life in New Zealand are found in a variety of settings. Fossil shellfish and other marine organisms occur in abundance in soft sedimentary rocks in a number of coastal areas of New Zealand, such as Bay of Plenty, Hawkes Bay, Wairarapa, Wanganui, Marlborough and North Canterbury. In the early and middle Pleistocene these areas were submerged beneath the sea, especially during warmer periods between ice ages, and only in the last few hundred thousand years have they been pushed up to become land.

Continued rapid uplift of New Zealand's mountains, together with the removal of forest cover in the ice ages, resulted in increased erosion. The gravel, sand and wind-blown dust accumulated in low-lying areas and formed the plains of Canterbury, Southland, Waimea, Horowhenua and Hauraki, to name just a few. Fossil pollens and spores in these sediments document the cyclical vegetation changes during the Quaternary ice ages. Fossil forests are occasionally found in these terrestrial deposits, or buried beneath lava flows or volcanic ash, in the central and northern North Island, where volcanoes were particularly active in the Quaternary.

The most significant Quaternary fossils in New Zealand are the bones of land birds, reptiles and amphibians that are found preserved in caves, swamps and sand dunes. Virtually all are younger than 40,000 years old (latest Pleistocene and Holocene). Also included are the young unfossilised bones of tuatara, moa and other birds found in the middens (rubbish heaps) of the prehistoric Maori.

The arrival of humans in the last 1000 years, with rats, dogs, fires, and more recently chainsaws and bulldozers, has brought another major episode of extinction, especially of New Zealand's unique terrestrial fauna. Whether the world-wide impact of human civilisation will be more severe than the extinctions at the end of the Permian or Cretaceous remains to be seen.

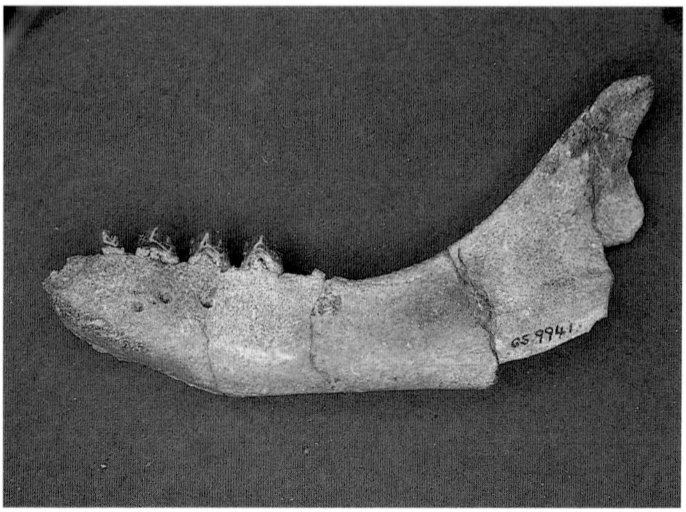

139. This 12cm jaw bone of a Ross seal (*Ommatophoca rossi*) was found fossilised in early Pleistocene (1.5 million year old) rocks near Napier. Today Ross seals live on pack ice around Antarctica. This fossil, together with those of a number of other cool water shells and crabs, provides convincing evidence of the cold temperatures around New Zealand in glacial episodes.

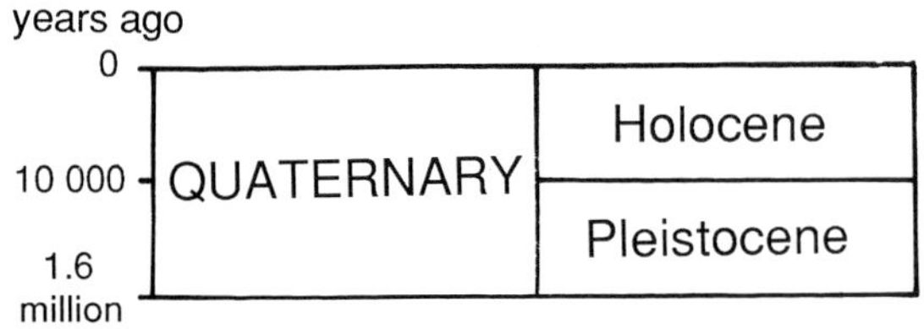

140. The Quaternary is divided into two periods.

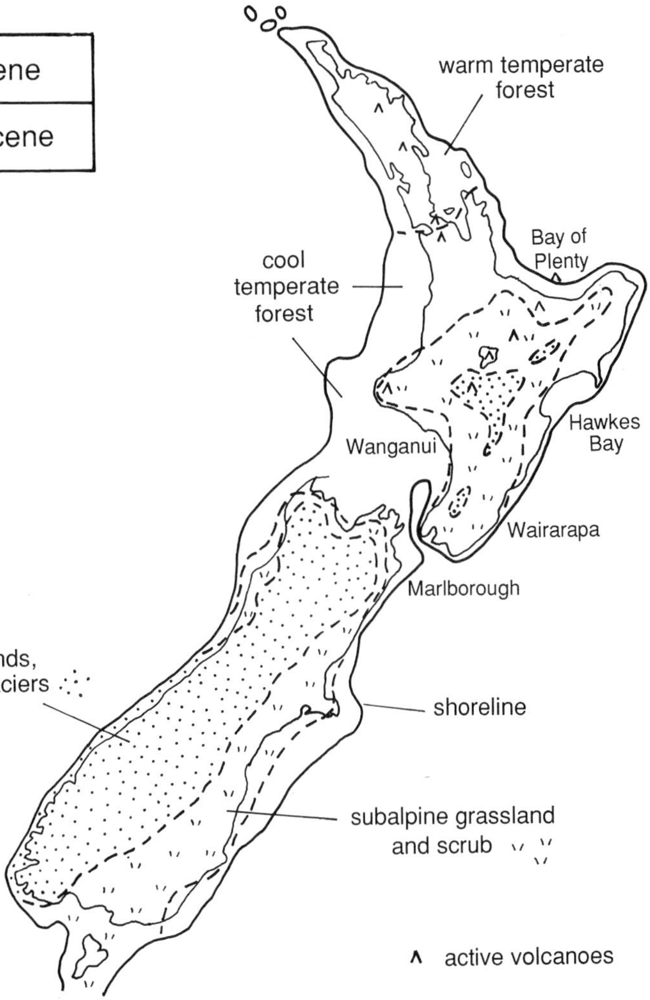

141. Map showing how New Zealand looked at the peak of the Last Ice Age (20,000 years ago) when sea level was about 150m lower than today and temperatures were much colder. (After Stevens and Fleming)

142. Fossil forests buried by volcanic eruptions or by moving sand dunes in the Quaternary are known in a number of places in the northern half of the North Island. In the far north of Northland, the stumps of a 30,000 year old fossil kauri forest can be seen at Lake Ohia, where they were preserved in a peat swamp behind moving sand dunes. Photo: Gordon Ell.

PENTAX

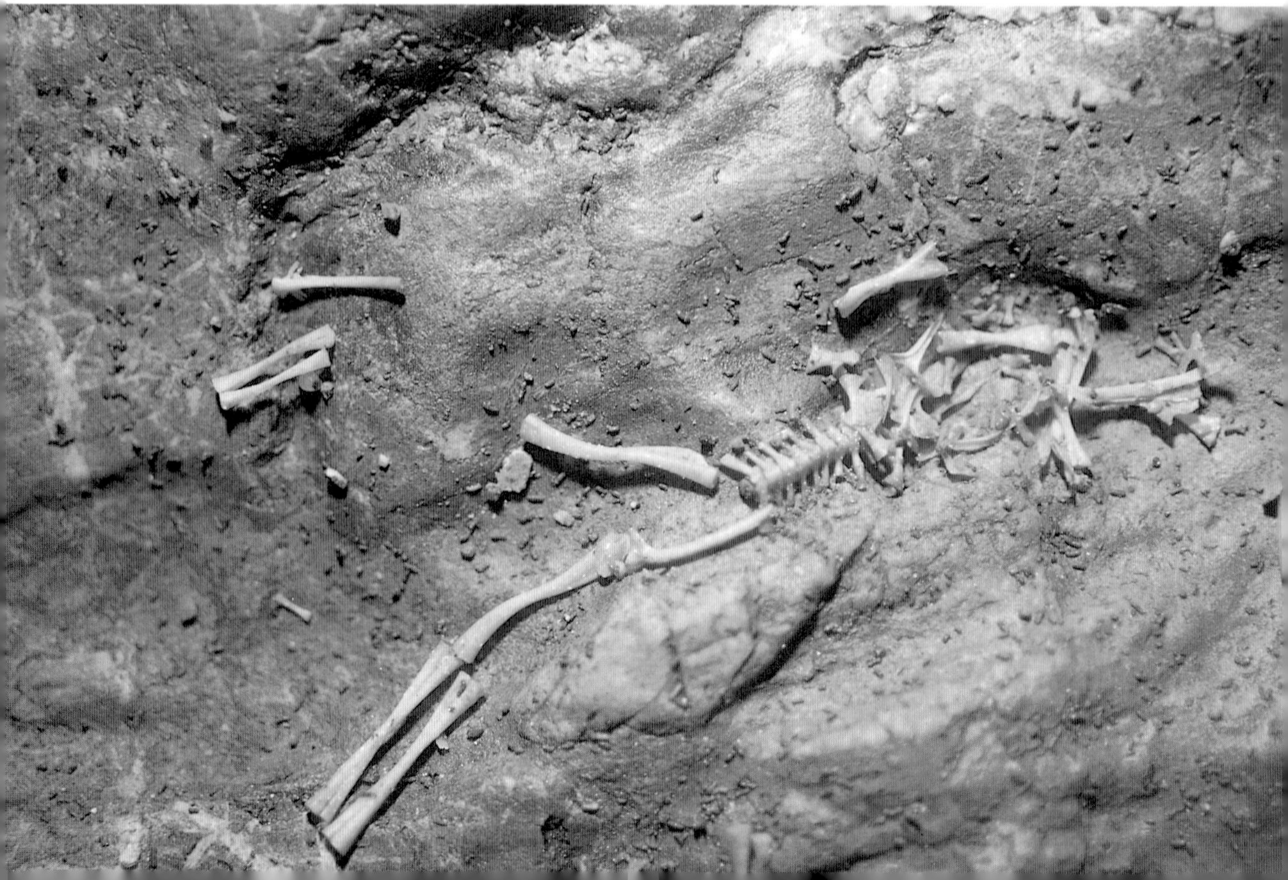

143. (Top left) Part of the subfossil skeleton of a large moa, *Pachyornis mappini*, together with several subfossil flax snail shells, *Placostylus ambagiosus*. They have been unearthed by moving wind blown sand in the dunes behind Tom Bowling Bay, North Cape. The dead moa and snails were buried 2000-3000 years ago by an advancing sand dune and remained preserved and hidden until recently exposed by the elements. Hundreds of similar bones of moas and other birds that have been preserved in dune sands in northern New Zealand have been found and studied in recent years. Photo: Phil Millener.

FROGS AND REPTILES

144. (Bottom left) The subfossil skeleton of the native frog, *Leiopelma hamiltoni*, in a cave in northwest Nelson. Subfossil bones of native frogs are quite common in cave deposits throughout New Zealand, although none are older than 10,000 years (Holocene). Today our unique, primitive native frogs consist of 3 species restricted to small areas in northern New Zealand and on Stephens Island, Cook Strait. Studies of their subfossil bones show that prior to the arrival of humans and rat predators, at least 6 species lived in New Zealand and they were widespread throughout both islands. Photo: Trevor Worthy.

Studies of subfossil bones of skinks, geckos and tuatara from caves and sand dunes also show that they were much more widely spread throughout the country prior to the introduction of predatory rats. The bones of at least one large extinct skink are known from Northland.

145. (Right)A paleontologist carefully excavates 30,000 year old subfossil moa bones buried in sediment on the floor of a cave at Oparara, north Westland. Photo: Trevor Worthy.

Birds

Bird bones are lightly constructed, often fragile and do not preserve well. As a result fossilised bird bones in New Zealand rocks are extremely rare. Apart from penguin bones, we have a fossil Oligocene albatross wishbone; a Pliocene seabird with tooth-like serrations on its bill; a handful of unstudied Miocene bones from Central Otago; and several late Pliocene and Pleistocene moa bones.

Our knowledge of New Zealand's endemic bird fauna, before the arrival of humans, comes from a number of rich finds of subfossil bones of late Quaternary age. Most are 1000-5000 years old but a few ancient cave deposits are up to 35,000 years old. The bones have been found in abundance, buried in sand dunes in northern Northland, or in caves especially around Waitomo, northwest Nelson and northern Westland.

146. The richest and best preserved assemblages of subfossil bird bones are found buried in scree and alluvium in ancient caves, such as here at Oparara in north Westland. These bones are from a small moa, *Megalapteryx didinus*, and an extinct flightless rail with an adzebill, *Aptornis otidiformes*. Photo: Trevor Worthy.

The lack of mammals or other predators in New Zealand allowed a highly specialised bird fauna to evolve. Flightlessness, giantism, low clutch size, and nocturnal behaviour had become common. These characteristics made them more susceptible to extinction when humans arrived, clearing the forest habitat, introducing predators and hunting the large birds. More than 40 percent of landbird species occurring 1000 years ago are now extinct. Our total knowledge of many comes only from their subfossil bones.

In addition to moas, other spectacular birds to have disappeared include the world's largest and most powerful eagle, Haast's eagle with a wing span close to 3m; a ferocious New Zealand goshawk; a large flightless goose; a flightless duck; a New Zealand crow; an owlet-nightjar; and a laughing owl. Their subfossil bones also illustrate how rich and varied were the birds of our prehistoric forests compared with today, when birds have been so decimated by rats and other introduced predators.

147. (At left) The largest moa, *Dinornis maximus*, stood 1.6m high to the rump and about 3m high overall with its neck outstretched. It would have weighed 200-300 kg and is the tallest bird known to have lived on earth. In life it probably walked around with its head and neck bent down somewhat, unlike the stance in this reconstructed subfossil skeleton. Photo: Alexander Turnbull Library.

Moas

Moas belong to an ancient group of flightless birds called ratites. Because ratites have representatives on all continents that once were part of Gondwana, it is thought their ancestors evolved there prior to Gondwana's breakup. The ancestor of the moas is thought to have come with New Zealand when it split away from Gondwana some 80 million years ago and gradually evolved into a number of different forms. Despite this supposedly long history in New Zealand, the oldest known fossil moa bone is only two million years old (late Pliocene) and almost all our knowledge of moas comes from subfossil bones and skeletons that are younger than 40,000 years, mostly 500-2000 years old.

Many thousands of subfossil moa bones have been found buried in sand dunes in northern Northland, in limestone caves, in swamp deposits and in prehistoric Maori middens. Last century enormous numbers of moa bones were found scattered over the ground surface, especially in Otago, Canterbury and Marlborough. The best complete skeletons, sometimes with intact gizzard contents, have mostly been excavated from swamp deposits in the eastern South Island. Several dry caves in Central Otago have yielded dried skin with feathers attached and claws and scales on the subfossil feet.

Studies suggest that when humans arrived about 1000 years ago, several hundred thousand moas lived in our primaeval forests eating green sprigs and woody twigs. They were hunted for food, harassed by native dogs, their forest habitat cleared, and their nests plundered by introduced rats, so that by 500-600 years ago they had virtually been wiped out.

148. Leg and foot bones of a large squat moa, *Pachyornis australis*, excavated in cave deposits at Oparara, north Westland. The large bone is 30 cm long. Because of the size and strength of their legs, leg bones are the most common subfossil remains of moas that are found. Photo: Trevor Worthy.

Current studies of their bones suggest that there were 11 species of moa in 6 genera. The largest, smaller was *Emeus*, about the size and shape of modern emus. The two smallest and most slender moas were *Megalapteryx* and *Anomalopteryx*, both about 1m high and 30kg in weight.

Kiwis are also ratites but they are so different from moas that they are thought to have evolved from an ancestor that flew into New Zealand in the early Tertiary.

Fossils and people

Modern society and the standard of living we enjoy relies heavily on the rock and mineral resources of the earth. Petrol, gas, coal, steel, aluminium, concrete, bitumen, fertiliser are just a few of these resources that contribute to our everyday lives.

To continue finding further reserves of these important minerals and rocks requires detailed studies and improved understanding of the geology of the earth. Fossils play a vital role in these studies, being used extensively to date the different strata and to indicate past environments and geographies. Nowhere are fossils more important than in the search for oil and gas. In New Zealand thousands of microfossil samples have been dated and used to better understand the sequence of rocks encountered in over 100 oil wells, drilled between Northland and the Great South Basin, in the last 30 years.

In some instances the valuable rock or mineral is actually composed of fossils. For example the most versatile and widely used rock on earth is limestone, which is formed of whole or broken up fossil shells, composed of calcite (calcium carbonate). Limestone is widely distributed and quarried throughout New Zealand and the world for a variety of purposes. Its largest uses are as the raw material for cement and agricultural lime. In both instances it is pulverised to a fine powder and cooked at very high temperature in a kiln. This drives off carbon dioxide, leaving calcium oxide powder, known as lime. Lime is used in steel

149-150. In the search for oil and gas, potential reservoirs deep in the earth are identified by geophysical methods. Then a well is drilled to test for hydrocarbons. As the drillbit cuts deeper and deeper into the earth, drilling mud is circulated down the hole. It then carries small chips of rock from the bottom, back up to the surface. Microfossils are extracted from these chips and their age used to determine progress towards the target. The age of the fossils is often valuable in determining structural complications. These may be critical to successful drilling or in determining the size of reserves encountered. Lower left: A drillship used for drilling exploration at sea. Lower right: A paleontologist studying microfossils down a microscope.

manufacture as a flux; to purify water; to treat sewage and industrial waste; as an ingredient in the manufacture of bleaches, pesticides, paint pigments and other chemicals; in sugar refining, brown paper production; as filler in rubber and paint; as one of the three main ingredients of bottle and window glass, and as a constituent of mortar and stucco, to name just a selection.

Limestone and marble (metamorphosed limestone), both of which are composed largely of fossil shells, have been widely used in New Zealand as a building stone in the past. Oamaru limestone and Takaka marble are still occasionally used in buildings.

Lignite and coal are also largely composed of fossil material, in this case fossilised plant matter. Coal is widely used in industry, such as the manufacture of steel at Glenbrook. Our coal and lignite fields in Waikato, north Taranaki, West Coast and Southland are New Zealand's largest fossil energy reserves.

Diatomite is another rock composed almost entirely of fossils. The rock is extremely porous and light because of the irregular shapes of the billions of microscopic diatom tests (plant plankton) that it is made of. When powdered, diatomite is a valued filter for water, wine, oil and other liquids. It is added to paint, as the angular diatom particles roughen the surface to deflect light and produce a flat finish. Diatomite's great value therefore rests on the strange and beautiful forms that one-celled plants made with silica, millions of years ago. Most diatomite used in New Zealand is imported, though we have deposits in the Waikato and near Oamaru.

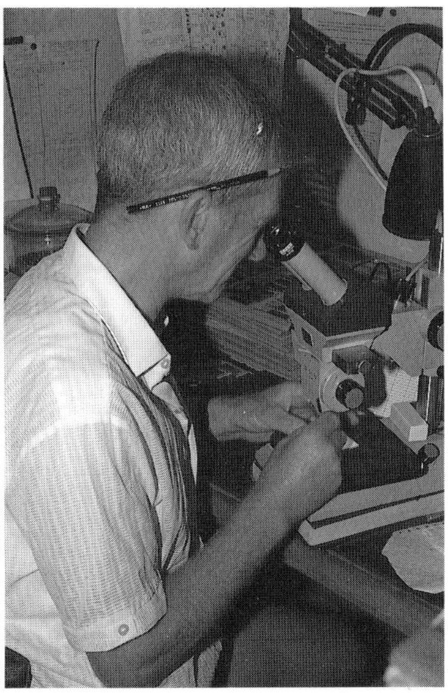

151. Takaka Marble on the top of Takaka Hill (northwest Nelson), is composed of 450 million year old fossil material. It is used today as decorative cladding panels (eg. in Parliament buildings) and as flooring or wall tiles.

152. Limestone is composed of whole or broken up fossil shells that accumulated on the sea floor as vast shell banks, millions of years ago. These deposits have since hardened and been uplifted to form land in many parts of New Zealand. They are now quarried extensively as a source of agricultural lime, for cement, building stone and a wide variety of other uses.

153. Coal, seen here in Waipuna opencast pit near Huntly, is composed of fossil plant material that accumulated in swamps millions of years ago. It is widely used in industry and constitutes New Zealand's largest fossil energy reserves.

Hints on collecting and preparing fossils

Finding fossils

First you need to find an exposure of rock that contains fossils. You may use this book, especially the section on where to find fossils in New Zealand, as a guide to the best areas. You may wish to join an amateur geology club, or rock and mineral club, and get more detailed advice from them. As you become more experienced and knowledgeable you will find geological maps, published by the Geological Survey, a valuable guide to finding suitable places for collecting.

Rock exposures are usually in river banks, stream beds, road cuttings, shore platforms, seacliffs and quarry faces. Great care should always be taken at the foot of cliffs as your hammering may cause rocks to fall from above. Also be very wary about climbing up unstable cliffs in search of fossils. Your safety is more important than any old fossil! Always ask permission before entering private property to search for fossils and remember that no collecting or hammering is allowed in National Parks and Scientific Reserves.

A few rocks (eg. certain limestones) consist almost entirely of shell-like fossils which can easily be seen. In other rocks, the fossils may be far less common and you need to carefully examine the surface in search of them. Sometimes they stand out from the surrounding rock because of a difference in colour or texture. Commonly only a small portion of fossil may be visible, the rest needs to be dug out and later exposed.

Sometimes a good place to find fossils is in the scree at the foot of exposures of softer rocks, as the fossils tend to wash out and accumulate there. The waste rock in coal mine tips is often worth searching for fossil leaves.

154. The majority of fossils in New Zealand can be collected together with the surrounding rock, using a hammer and cold chisel. A few rocks are so soft that a pocket knife will prize the fossils out. Some limestones and older rocks are so hard that a heavy hammer is needed to break off pieces, hopefully containing the fossil.

Collecting

A fossil collector should not set off without a geological hammer, cold chisel, pack, pocket knife, magnifying glass, notebook and pencil. A spirit-based felt pen is useful for numbering samples or plastic bags; newspaper is essential for wrapping up specimens to protect them; and a small box lined with cotton wool is a good idea for carrying delicate or small fossils. Safety glasses or sunglasses are important to protect your eyes from flying chips when collecting in hard rocks. Where possible carry a map (topographic or geological) and accurately record the locality and kind of rock you collected each specimen from. Many major New Zealand fossil finds have been made by dedicated amateurs and it is important to record details of the find for your own reference and in case the fossil turns out to be important.

Methods of getting fossils out of rocks in the field depend on the hardness of the rock. Some Quaternary and Tertiary rocks are soft enough that fossils can be picked out using a pocketknife. Other rocks require a hammer and cold chisel and an element of luck, as you patiently chip away around the fossil until a portion of rock, hopefully containing the whole specimen, is freed. As you chip around the fossil, slant the chisel away from it, to help prevent flake damage.

Some of our Paleozoic and Mesozoic marbles and greywackes are too hard to chip out fossils with a chisel. In these cases, you will have to make do with breaking off sharp edges of the rock or with splitting off pieces along natural fractures.

155. As your collection grows, you may wish to display some of your fossils or to store them in drawers, neatly labelled and boxed, like these fossil scallops (*Pecten*).

Cleaning and curating fossils

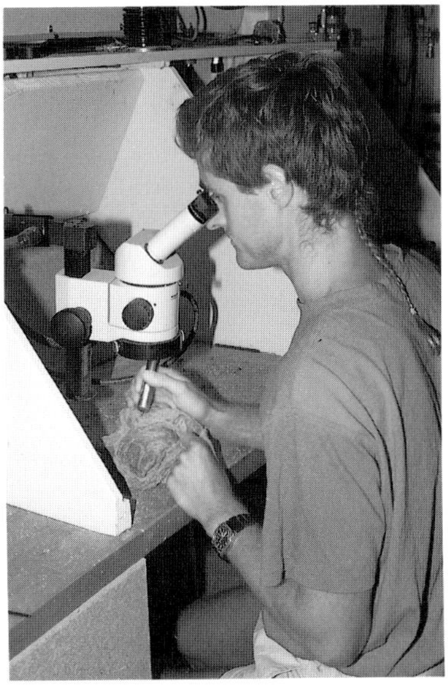

156. A paleontologist uses a small compressed air drill in the laboratory to remove fragments of rock that surround a fossil bivalve.

Having got your collection home, you may want to expose more of your fossil by removing the surrounding rock using a small cold chisel and hammer, or sharp dental picks if the rock is soft enough. Usually work along the visible edge of the fossil and chip the rock off in small flakes. Cleaned specimens that are powdery can be protected by spraying or gently brushing on a thin coat of clear polyurethane lacquer. A coat of this is also useful to prevent some soft mudstones and fossils from drying out and fragmenting into little pieces.

Having cleaned your fossils, you may wish to display your best specimens. You could mount them on low pedestals, or keep them in small cardboard or plastic boxes arranged neatly in drawers, with their labels. You should keep track of your fossils by recording locality and identification details in a notebook, card index or computer file. Reference numbers can be written in Indian ink directly on the rock or fossil, or on a small spot of white paint or snowpake.

Professionals and a few dedicated amateurs in New Zealand use more expensive mechanical and chemical methods to clean and expose their fossils. Fossils can be extracted from even our hardest rocks by dissolving away the surrounding rock over a period of months using dilute acetic acid. Replicas of important specimens are sometimes produced, using rubber latex to make a mould, and plaster of Paris to make the cast.

157. When the original fossil shell has been dissolved, the natural mould that is preserved in the rock may be coated with rubber latex. When set, the rubber is peeled off revealing a replica of the shell and its fine structures and ornament, ready for study.

Where to see and find fossils

Numbers in brackets at the end of each entry refer to photographs in this book of the site, or fossils found there.
* An asterisk indicates private property requiring permission to enter.

NORTH ISLAND

1. Lake Ohia fossil kauri forest, Doubtless Bay. Visitor area 1 km along road to Whatuwhiwhi from Hwy 10. Stumps of late Quaternary forest preserved in partly drained peat swamp. [142]
2. Winstones Otaika Quarry, Whangarei. At top of west end of quarry, in greensand and limestone above greywacke. Eocene calcareous algae and large bivalves.*
3. Pakaurangi Point, SE Hukatere Peninsula, Kaipara. 4km walk along coast (low tide) from Tinopai. Abundant early Miocene snails, bivalves and variety of other fossils in cliffs.
4. Mathesons Bay, Leigh. In conglomerate and sandstone rocks at SE end of beach. Giant fossil oysters, calcareous algae and rarer lampshells and corals. Early Miocene age. [115]
5. Black Rock tree moulds, Takapuna. In sea cliffs between ends of Brett and O'Neills Avenues. Moulds of 10-50 cm diameter tree trunks in lava flow from Lake Pupuke volcano. [17]
6. Motutapu Island, Waitemata Harbour. At foot of seacliffs, 1.5km walk (low tide) north of Islington Bay. Large Miocene barnacle plates in sandstone and grit beds. [133]
7. Ihumatao fossil forest, Mangere. At end of Renton Rd, 2km west of airport. In foreshore and low cliffs. Trunks and stumps of 30,000 year old kauri fossil forest buried by volcanic ash from Ihumatao volcano.
8. Coromandel Range silicified wood. Found in a number of places — eg. Amodeo Bay, Coroglen, Rangihau V., Kauaeranga V., east Fletchers Bay, eroding out of late Tertiary volcanic sequences. Permit to collect required in Coromandel Forest Park.
9. Moeweka Quarry, Port Waikato. Disused quarry 400m west of Ponganui Rd, up Moeweka Stream. Rich fauna of bivalves (esp. *Inoceramus*) and belemnites, in Jurassic greywacke.*
10. Huriwai plant beds, Port Waikato. In low cliffs at north end of Huriwai Beach. 5km S of Port Waikato. 3km walk from coast road. Well preserved Jurassic ferns and leaves. [104]*
11. Whakapirau Rd, S. Kawhia. In numerous road cuttings along 8km stretch. Well preserved Jurassic bivalves and less common belemnites, lampshells and ammonites.
12. Kiritehere Coast. In seacliffs and shore platform along 3km stretch southward from Kiritehere Beach. Well preserved Triassic clams, mussels, lampshells and rare snails. [85, 89]
13. Airedale reef, Waitara. Headland reef, 1.5km east of Waitara R. mouth. In-situ trunks and roots of buried Quaternary fossil forest.
14. Nukumaru Beach. 15km NW of Wanganui. Road turnoff at Waitotara. In seacliffs (low tide access). Rich, well preserved Pliocene bivalves and snails in soft sandstone.
15. Castlecliff, Wanganui. 1.5-2km walk (low tide) along beach, north of end of Karaka St. Abundant Pleistocene bivalves and snails, in soft sandstone forming sea cliffs. [138]
16. Titahi Bay fossil forest, Wellington. In intertidal reef in middle of bay. Remains of 50,000 year old forest.
17. Matata, Bay of Plenty. In bank at mouth of small stream gorge, 400m west of Hereperu Rd jctn., 100m S of Hwy 2. Rich Pleistocene bivalves and snails in soft siltstone.

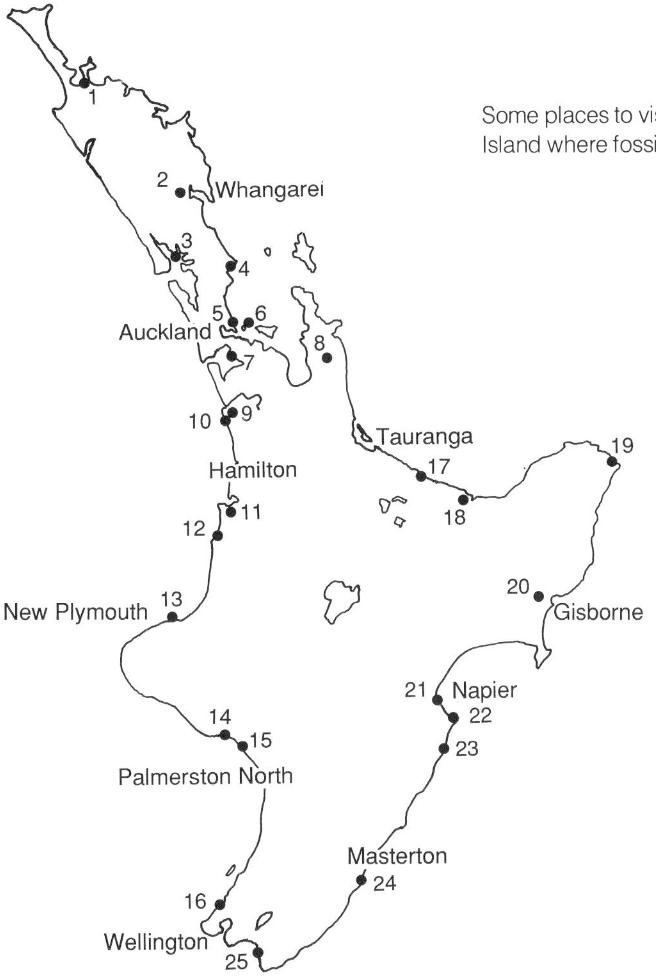

Some places to visit in the North Island where fossils can be seen.

18. Ohiwa Harbour, Bay of Plenty. In cutting on Ohope-Kutarere Rd, 4km S of Ohope Beach, west shore of harbour. Abundant Pleistocene bivalves and snails in soft siltstone.
19. Te Araroa, East Cape. 1km east of Te Araroa in road cuts and shore platform. Well preserved late Miocene bivalves, snails and trace fossils.
20. Patutahi and Waerengaokuri Quarries, Gisborne. 20km and 25km inland from Gisborne on Wharekopae Rd and Hwy 36. Rich, well preserved late Miocene bivalves (scallops) in limestone.*
21. Scinde Island, Napier. In cliffs at corner of Hyderabad and Battery Rds; and behind carpark opposite wharves. Pliocene bivalves (scallops, oysters) and trace fossils in limestone.
22. Cape Kidnappers, Hawkes Bay. 1-6km east of Clifton, as far as Black Reef. In sea-cliffs (low tide access). Scattered horizons rich in Pleistocene cockles and other bivalves. [12]
23. Waimarama, Hawkes Bay. Half hour walk (low tide) south from end of Waimarama Beach road. Just north of Red Island. Fallen blocks rich in excellent Cretaceous bivalves (*Inoceramus*).
24. Castlepoint, Wairarapa. S end of beach in fishing reef and lighthouse point. Abundant lampshells and bivalves (scallops) in Pliocene shelly limestone.
25. Putangirua and Hurupi Streams, Palliser Bay. In banks and bed of stream and track to Putangirua Pinnacles and in lower part of Hurupi Stream, 1km to N. Common Miocene bivalves and snails.

SOUTH ISLAND

26. Max's Bush, Wairoa R, Richmond. In road cuttings for 1.5km upstream from picnic area, east side of Wairoa Gorge mouth. Greywacke with beds rich in Triassic bivalves and lampshells.

27. Baton River, inland Motueka. In river bed and banks, beside swing bridge, 4km up from end of Baton Valley Rd. Moulds and casts of Devonian lampshells, bivalves and rare corals, sea lilies and trilobites in greywacke. [67]

28. Tarakohe Quarry, Golden Bay. In siltstone above limestone in cement quarry, 1.5km NE of Pohara. Abundant early Miocene snails, bivalves, corals and many other fossils. [122] *

29. Trilobite Rock, Cobb Valley. At head of Cobb Reservoir. Cambrian trilobite heads and tails, best seen in weathered faces. NZ's oldest fossils. Protected, no hammering. [62, 72-74]

30. Chaffey Stream, Cobb Valley. 6km walk up Cobb Valley track from road end. Ordovician graptolites found on bedding planes of black slate boulders in stream bed. [75]

31. Mangles Valley, Murchison. In roadcut on Mangles-Tutaki Rd, 400m from jctn with Nelson-Murchison Hwy. Rich late Miocene leaf impressions in mudstone within conglomerate sequence. [134]

32. Reefton Devonian fossils. In roadcut on road to Lewis Pass, 6.5km SE of Reefton. Moulds and casts of bivalves, lampshells, corals and sea lilies in hard greywacke and limestone. [5, 28, 66, 78, 79, 81]

33. Strongman Mine dump, Greymouth. 1.5km west of Hwy 6, 6km N of Runanga. In dump below mine. Cretaceous ferns and flowering plant leaf impressions in mudstone. *

34. Oaro, Kaikoura Coast. 1.5-2km walk along coast track beside railway line, S from Oaro. Common belemnites, bivalves, reptile bones and silicified wood in late Cretaceous sandstone.

35. Motunau Beach, North Canterbury. In cliffs, boulders and shore platform S of Motunau. Spider crabs in concretions, rare whale and penguin bones and other fossils. Pliocene. [31]

36. Weka Pass, Waipara. In Weka Ck and cuttings on Culverden Rd through Pass. Abundant Miocene lampshells, bivalves and snails in sandstone and limestone sequence of rocks. *

37. Glenafric, Waipara. In seacliffs, 1km walk from end of Mt Cass Rd, E of Waipara. Miocene crabs, bivalves, nautiloids, corals and whale bones in concretions in siltstone. *

38. Curiosity Shop, Rakaia Gorge. In old lime quarry on NE bank of Rakaia R, 6km S of Windwhistle. Rich, well preserved late Oligocene bivalves, lampshells, sea eggs, snails and shark teeth. *

39. Trig Z, Waitaki Valley. 14km W of Duntroon, 300m NW of Gards Rd, 1km S of Hwy 83. Abundant early Miocene bivalve, lampshell and snail fossils in upper part of limestone bluffs. *

40. Livingstone plant fossils, Waitaki V. In hills, 1.5km W of Livingstone, 16km S of Duntroon. Good early Tertiary leaf impressions in mudstone within conglomerate sequence. *

41. Old Rifle Butts, S side Cape Wanbrow. 1km walk N along beach from where coast road S from Oamaru meets sea. Common early Miocene lampshells in greensand in cliffs.

42. Burnside Cement Quarry, Dunedin. N of railway line, Green Island. Rich lampshells, bivalves, sea eggs and shark teeth in Oligocene limestone and greensand. *

43. Tunnel Beach, Dunedin. 600m walk from end of Hillhead Rd, W of St Clair. Scattered Miocene lampshells and bivalves in cliffs.

44. Fairfield Quarry pit, S Dunedin. NW side of Hwy 1, 1km E of Fairfield. Common late Cretaceous and Paleocene bivalves, lampshells, some nautiloids and ammonites. *

45. Landslip Hill, E of Gore. Roadside and paddock quartzite boulders, in hills just N of Pukerau, 10km E of Gore. Moulds and impressions of Miocene leaves and seeds. *

46. Arthurton Quarry. On W bank of Waipahi R, Arthurton, 15km E of Gore. Excellent large Permian bivalves (*Atomodesma*) in hard limestone. *

Some places to visit in the South Island where fossils can be seen.

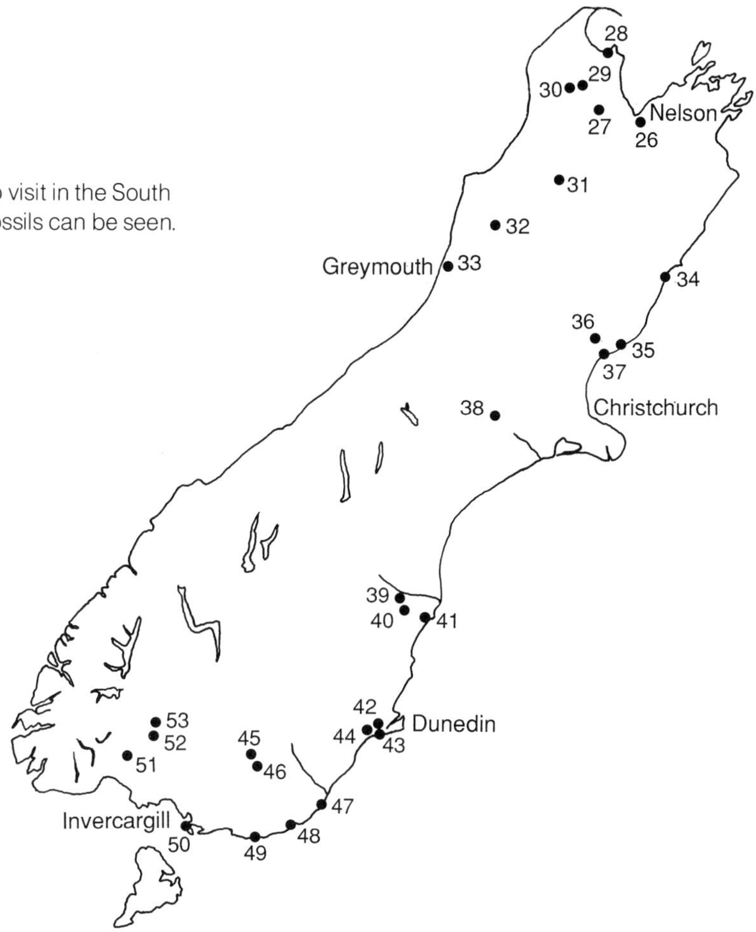

47. Nugget Point area, Catlins Coast. In coastal cliffs, shore platform and roadcuts between Campbell and Nugget Pts and in Roaring Bay to S. Rich Triassic bivalves and lampshells. [88, 90]

48. Papatowai, Catlins Coast. 0.5-2km S around coast from Papatowai. In cliffs and shore platform. Rich early Jurassic bivalves and lampshells in greywacke.

49. Curio Bay fossil forest, Waikawa. In cliffs and shore platform, 5km S of Waikawa. In-situ stumps plus logs and leaves of Jurassic forest. No hammering, Scientific Reserve.

50. Mokomoko Inlet, Bluff. In shore platform, E side of entrance to Mokomoko Inlet. Access from end of Stanley Township Rd. Well preserved Permian trace fossils.

51. Clifden. 12km N of Tuatapere. In river bank for 2km W of old suspension bridge beside Hwy 99. Abundant, well preserved Miocene bivalves, snails and lampshells. Scientific Reserve. [82, 124, 130]

52. Ohai opencast mine dump. Just N of Hwy 96 at Ohai. Cretaceous leaf impressions and freshwater mussels in mudstone. *

53. Productus Creek, Ohai. In bed of creek, 10km on farm road through Beaumont Station, 5km N of Ohai. Permian lampshells, bivalves and snails. *

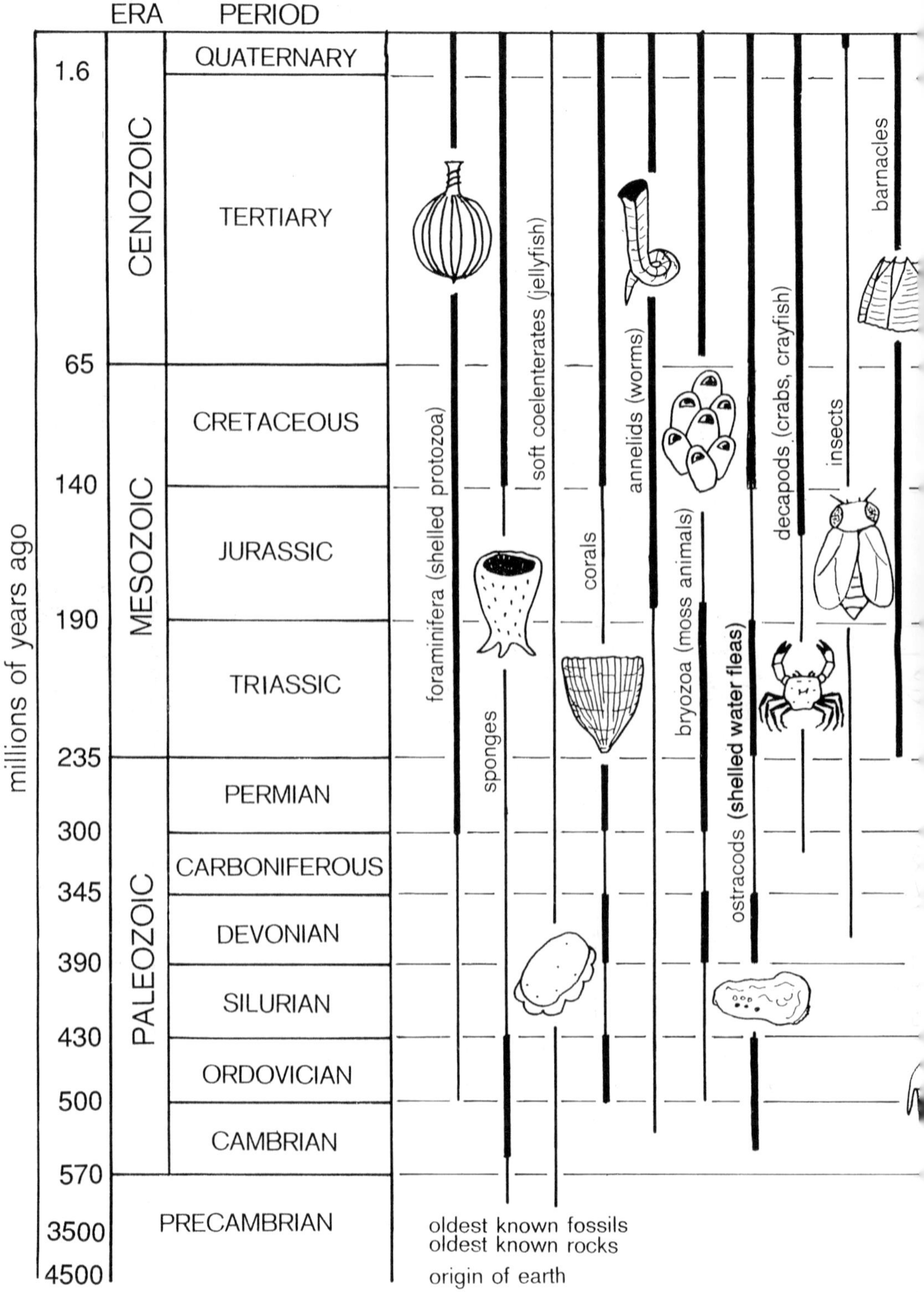
ERA
PERIOD
millions of years ago
1.6
65
140
190
235
300
345
390
430
500
570
3500
4500
CENOZOIC
MESOZOIC
PALEOZOIC
QUATERNARY
TERTIARY
CRETACEOUS
JURASSIC
TRIASSIC
PERMIAN
CARBONIFEROUS
DEVONIAN
SILURIAN
ORDOVICIAN
CAMBRIAN
PRECAMBRIAN
foraminifera (shelled protozoa)
sponges
soft coelenterates (jellyfish)
corals
annelids (worms)
bryozoa (moss animals)
ostracods (shelled water fleas)
decapods (crabs, crayfish)
insects
barnacles
oldest known fossils
oldest known rocks
origin of earth

TIME CHARTS FOR FOSSIL GROUPS

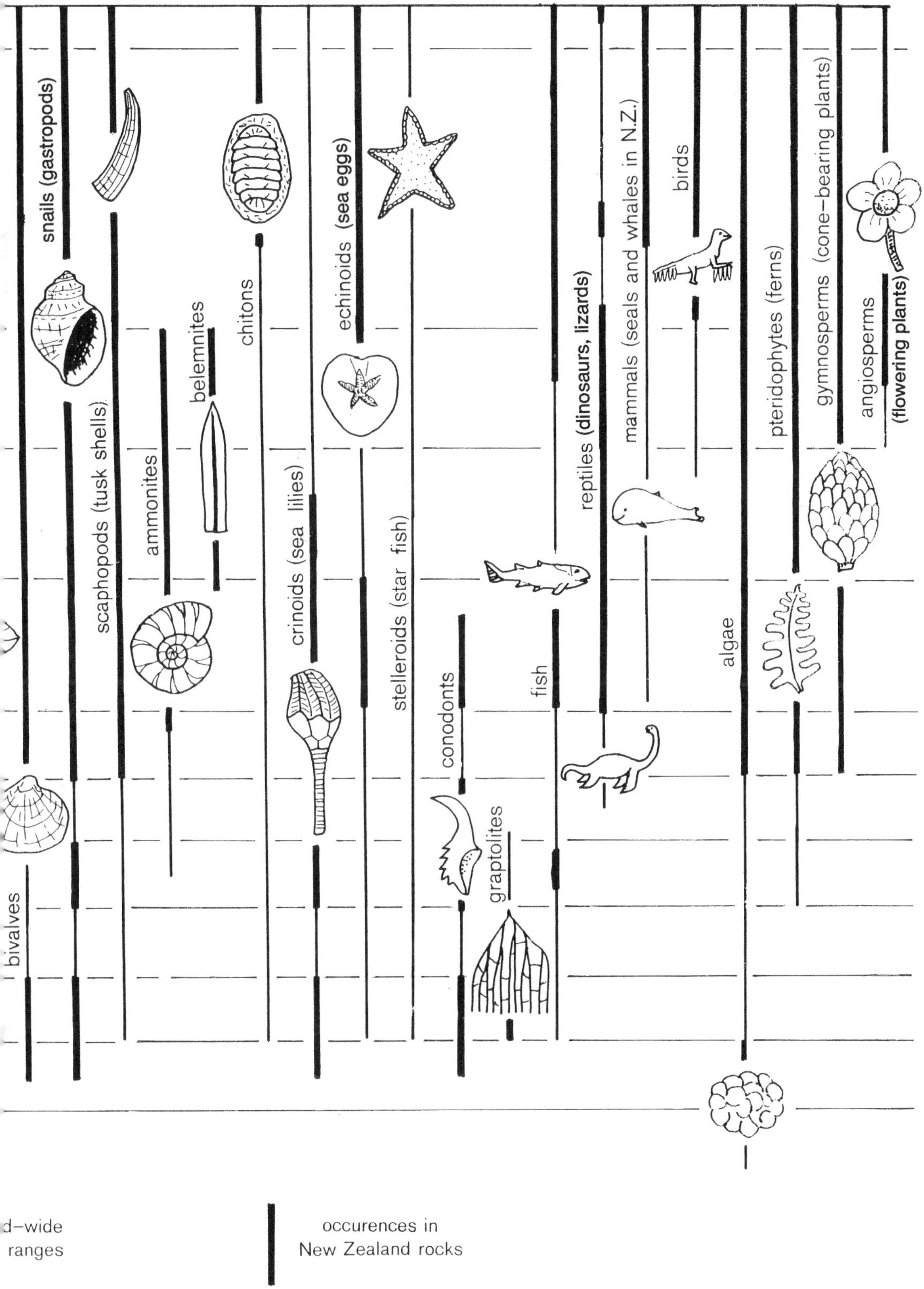

Further reading

Prehistoric New Zealand (Heinemann Reed, 1988, 128p) by G. Stevens, M. McGlone, B. McCulloch, and V. Ward.

Field guide to New Zealand geology (Reed Methuen, 1985, 226p) by J. Thornton.

Lands in collision. Discovering New Zealand's past geography (Science Information Publishing Centre, 1985, 129p) by G. Stevens.

New Zealand adrift. The theory of continental drift in a New Zealand setting (Reed, 1980, 442p) by G. Stevens.

Cobb Valley: a geological guide (Geological Society of New Zealand Guidebook No 6, 1984, 48p) by R. Cooper.

The geological history of New Zealand and its life (Auckland University Press, 1979, 141p) by C. Fleming.

Illustrations of New Zealand fossils (NZ Geological Survey Handbook, 1981, 109p) by I. Speden and I. Keyes.

Manual of New Zealand Permian to Pleistocene foraminiferal Biostratigraphy (NZ Geological Survey Paleontological Bulletin 56, 1989, 175p) by N. Hornibrook, R. Brazier and C. Strong.

Cenozoic Mollusca of New Zealand (NZ Geological Survey Paleontological Bulletin 58, 1990) by A. Beu, P. Maxwell and R. Brazier.

Te Moa. The life and death of New Zealand's unique bird (Nikau Press, 1987, 64p) by B. Brewster.

GLOSSARY

alluvial — transported or deposited by rivers.

Alpine Fault — major fracture through the South Island, extending from Blenheim to Milford Sound. The east side has moved 500km south with respect to the west side.

ammonite, ammonoid — extinct group of squid that lived in chambered shells. The internal walls between chambers are wavy or intricately frilled. Common fossils in Paleozoic and Mesozoic rocks.

anaerobic — without oxygen.

andesite — a grey volcanic rock.

angiosperm — a flowering plant.

araucarian — group of primitive conifer trees that includes kauri and *Araucaria.*

belemnite — extinct group of squid that had an internal bullet-shaped hard part that is commonly fossilised in Mesozoic rocks.

brachiopod — lampshell. A sedentary, marine animal resembling a Roman oil lamp. Consists of two shells, of unequal size, often attached to the bottom by a muscular stalk.

bryozoa — marine ''moss animals'', consisting of polyps, each with a horny or calcareous cup that join together to form erect or encrusting colonies.

calcite — the white or cream mineral, calcium carbonate, $CaCO_3$. The main component of shells and limestone.

calcareous, calcified — made of calcite.

cast — form produced by casting inside a mould; eg. if mud hardens inside a shell and the shell dissolves away, an internal cast remains.

chitin — horny, organic substance that forms the exterior skeleton of crabs, insects, trilobites and other arthropods.

chiton — primitive marine shellfish (mollusc) with shell composed of a number of calcareous plates set in a leathery girdle. Plates are occasionally fossilised.

clutch size — set of eggs laid and hatched in a brood.

coccolith — microscopic, marine algae that live in the plankton and have their cell walls covered in minute calcareous plates that are commonly preserved as fossils in deep water sediments.

concretion — a hard, usually rounded mass of stone that forms within a sedimentary rock by the precipitation and accumulation of minerals, often around a nucleus, which may be a fossil.

conglomerate — a sedimentary rock containing rounded cobbles and pebbles derived by erosion of older rocks.

conifer — group of cone-bearing and pollen-producing trees, commonly known as pines.

conodont — microscopic teeth of small extinct sucker eels. Commonly fossilised in Paleozoic and Triassic rocks.

cycad — a primitive group of small, palm-like, cone-bearing trees.

diatom — microscopic algae with delicately sculptured, siliceous cell walls. Frequent in fresh and sea water and commonly preserved as minute fossils.

diatomite — sedimentary rock composed largely of siliceous diatom skeletons.

echinoderm — group of ''spiny-skinned'' marine animals that include sea eggs, sea lilies, starfish and sea cucumbers.

endemic — native; limited to one place or country.

foraminifera — microscopic, marine, one-celled organisms (protozoa) that have shells which are commonly preserved as minute fossils.

fossil — any remains or trace of an ancient animal or plant preserved in rock.

gingko — a primitive group of deciduous, gymnosperm trees, having fan-shaped leaves.

glacial episode — a cold period during the Ice Ages when ice sheets covered large areas of land, and sea level dropped to 50-150m lower than today.

Gondwana — name given to the ''super-continent'' that existed in the southern hemisphere prior to its breakup, about 100 million years ago.

gorgonian — coral-lke marine animals with long, slender, branching colonies and solid, internal, rod-shaped, skeletal pieces that are sometimes fossilised.

graptolite — small, extinct, colonial animals that lived in the plankton and consisted of one or more branches of polyps. Found fossilised in Paleozoic rocks.

greywacke — a hard, compacted, grey sandstone that forms the main axial ranges of most of New Zealand.

gymnosperm — plants with seeds but no flowers, and usually bearing cones.

hydroid — small, fresh water or marine, sea anemone-like animals with a fleshy stem and tentacles.

ichthyosaur — extinct marine dinosaur or sea dragon, sometimes called fish lizards and resembling large dolphins. Found fossil in Mesozoic rocks.

interglacial episode — a warm period (similar to our present climate) between two colder glacial episodes during the Ice Ages.

lignite — brown coal. Soft low-rank coal.

limestone — a sedimentary rock containing more than 50% carbonate, usually calcite. Mostly composed of whole or broken calcareous fossils.

mammal — vertebrate group characterised by presence of hair and secretion of milk for its young.

marble — a recrystallised limestone metamorphosed at high temperature and pressure.

marine — in the sea or sea-dwelling.

Mesozoic — age of reptiles, between 235 and 65 million years ago.

metamorphosed — greatly changed through the action of enormous heat and high pressures, usually accompanying deep burial.

midden — rubbish heap.

mosasaur — extinct marine dinosaur or sea dragon, sometimes called sea lizards and resembling modern monitor lizards. Found fossil in Mesozoic rocks.

mould — the hollow template or shape of a fossil that is left in the rock when a fossil dissolves away.

mudstone — a sedimentary rock composed of fine (mud size) particles.

nautiloid — a primitive group of squid that lived in chambered shells and are occasionally found fossilised. The internal walls between chambers are smooth and simple. Pearly Nautilus is the only living form.

ooze — mud that accumulates on the floor of the deep ocean from particles and plankton that slowly settle out of the sea water.

operculum — the calcareous or chitinous "door" that seals the opening of snail shells when the animal is inside.

ossicle — a small disc-like plate that forms part of the skeleton of sea lilies.

otolith — microscopic, calcareous bone in the inner ear of fish and other vertebrates.

paleontology — the study of fossils.

paleontologist — a person who studies fossils.

Paleozoic — age of invertebrates, fish and amphibians, between 570 and 235 million years ago.

peripatus — small, primitive, worm-shaped animal with legs and tentacles; intermediate between round worms and insects.

planispiral — coiled in one plane.

plankton — small plants and animals that live floating in water.

plesiosaur — extinct marine dinosaur or sea dragon, sometimes called sea serpents and similar to Loch Ness monster. Found fossil in Mesozoic rocks.

podocarp — group of conifer trees with winged pollen that includes totara, rimu and tanekaha.

polyp — a small sedentary animal with a fleshy stem and terminal tentacles around the mouth.

proteas — group of dicotyledonous flowering plants that includes the well-known garden shrubs.

Quaternary — period of time spanning most of the Ice Ages, between 1.6 million years ago and the present day.

radiolaria — microscopic, planktonic, one-celled, marine organisms (Protozoa) with siliceous shells. Often fossilised in deep water sediments.

sandstone — a sedimentary rock composed of sand particles.

schist — metamorphosed rock characterised by strong alignment of platy crystals.

sea lily — crinoid. A bottom-dwelling, marine animal with a calcareous stalk and flower-shaped cup, that is fossilised whole or broken up.

sea pen — sedentary, coral-like animals with feathery pen-shaped colonies that live in sediment on the sea floor.

silica — a colourless or white mineral made of silicon dioxide, SiO_2. Forms include quartz, opal and chert.

siliceous — made of silica.

skeletal — hard parts of an organism.

slate — low-grade metamorphosed mudstone with well-developed cleavage, but little recrystallisation.

subduction — the process in which the edge of one of the earth's crustal plates is forced down into the mantle beneath the edge of an adjacent plate.

subfossil — any remains of a plant or animal that has been preserved, but not yet fossilised in rock.

terrestrial — land-dwelling or land-based.

Tertiary — age of mammals, between 65 and 1.6 million years ago.

trilobite — extinct group of marine animals resembling large sea lice. Found fossilised in Paleozoic rocks.

Other books by Bruce Hayward include:

- *Kauri Timber Dams* (with John Diamond)

- *Kauaeranga Kauri*

- *Ancient Undersea Volcanoes: A guide to the geological formations at Muriwai, west Auckland*

- *The Maori history and legends of the Waitakere Ranges* (with John Diamond)

- *Cinemas of Auckland, 1896-1979* (with Selwyn Hayward)

- *Waitakere Kauri* (with John Diamond)

- *Kauri gum and the gumdiggers*

- *Granite and Marble: A guide to building stones in New Zealand*

- *Landforms and geological features: A case for preservation* (with Beth Vaughan and Jack McConchie)

The author

Dr Hayward is a paleontologist (studies fossils) at the New Zealand Geological Survey (DSIR), Lower Hutt, where he specialises in the study of fossil Tertiary foraminifera (microscopic, shelled, marine protozoa) and their use in determining past environments.

His broad interests in the natural and human history of New Zealand have resulted in prolific scientific and popular publications on fossils, geology, earth science conservation, marine ecology, lichens, prehistoric archaeology, kauri timber and gum industries and the natural history of our offshore islands.